WHEN Biometrics FAIL

Shoshana Amielle Magnet

WHEN Biometrics FAIL

Gender, Race, and the Technology of Identity

Duke University Press Durham and London 2011

Printed in the United States of America on
acid-free paper ∞

Designed by Heather Hensley

Typeset in Scala by Keystone Typesetting, Inc.

Library of Congress Cataloging-in-Publication
Data appear on the last printed page of this book.

This book is dedicated to my mother,
Sanda Rodgers, and my partner, Robert Smith?

CONTENTS

ACKNOWLEDGMENTS

The first draft of what became this book was started when I was a PhD student at the University of Illinois at Urbana-Champaign. I was lucky enough to be surrounded by scholars and friends who provided rich critical race feminist analysis that was characteristic of the Institute of Communications Research, where I was in residence. The help I received from members of the CU Feminist Collective in formulating this project cannot be overestimated. Aisha Durham first gave me the idea of writing about biometric technologies during a coffee session, when she suggested I take a look at their use in the regulation of welfare recipients. Her incredible insight, brilliance, and generosity as a scholar informed this book from the start. Without my study buddy, Celiany Rivera Velàzquez, I never would have made it through the writing stage of the draft that became this book. Celiany always pushed my theoretical boundaries. Her reflections on the complexities of hybridity, racialization, and queer feminisms were invaluable, and she always had an hour (or four) to bring her insight to bear on my project. I also remain incredibly grateful for the day that Himika Bhattacharya decided to join the Institute. Her dazzling intellect, inspired comments, and vast knowledge of intersectional feminist theory were incredibly important to my understanding of biometric technologies. All were fabulous for their theoretical help and for lending me their shoulders to cry on (which I did with regularity). Before I even arrived at the Institute a phone discussion with Jillian Baez while I was deciding where to go to graduate school was key. Jillian and I began at the Institute in the same year. Her razor-sharp mind pushed my own thinking. (And her hilarious sense of humor didn't hurt either!) Carolyn Randolph was a later addition to the collective, but her incisive comments and her organization, with Aisha Durham, of the Women of Color Feminisms reading group, were essential to my development as a scholar. No discussion of the CU Feminist Collective would be complete without a mention of Joan Chan. Joan has the kind of mind that makes tackling complex theory look easy, and her commitment to combating inequality remains an inspiration to me. Huge thanks also to Rachel Dubrofsky; her kindness, generosity, and insight as a feminist scholar and mentor have been and continue to be incredible. I cannot thank Amy Hasinoff enough for her help in thinking through some of the toughest problems I faced while writing, not to mention her incredible kindness and her cupcake baking that so often fueled my writing process! A big thanks to those who helped me settle in at Champaign in those first few years. Kumi Silva and David Monje were incredibly kind to me; had they not hosted my first visit to Champaign, I'm not sure that I ever would have chosen to go to UIUC, nor would I have made it through without them, Kevin Dolan, Grant Kien, Diem-My Bui, or Susan Harewood. Sasha Mobley needs a mention here too. She is one of those rare scholars who have both depth and breadth, and I learned much from talking with her. I must also give credit to Andrea Ray for her incredible support and administrative assistance at the Institute. A huge thanks also to C. L. Cole, John Nerone, and Steve Jones. Thanks to John for teaching me the history of communications and technology. Steve gently showed me how and where my work needed to engage with the Canadian School of Communications, which made my project much the richer. Cole also was wonderful from start to finish; she repeatedly steered me in new directions for thinking about technology and the body. Thanks also to Lisa Nakamura, whose work on race, racism, and technology inspired me from the early days of my undergraduate thesis. I feel incredibly lucky to have had the opportunity to have her insights into my project; her comments on the draft were instrumental to shaping my book and also have offered exciting new possibilities for future work.

Of course the greatest thanks must be extended to Paula Treichler and Kent Ono. Their absolutely brilliant comments on countless drafts, their support of my work through thick and thin, their continual mentoring, responding at times to daily phone calls, and their sheer intellectual edge and integrity were simply unbelievable. They also usefully modeled how to be brave academics in a time when such models are few and far between. Words cannot completely express my intellectual and scholarly debt to them. Also thanks to Paula's partner, Cary Nelson, our resident "poetry agitator." Hanging out at Paula's and Cary's house made me feel as if I had finally made it into the cool group. Cary was incredibly kind to me, and even though I wasn't his student he offered me much-needed advice and talked me down from the ledge more than once (especially after a particularly memorable job interview!).

Kelly Gates's willingness to advise me on the pleasures and pitfalls of writing and research was equally generous and constant. Her comments on an early version of the biometrics and welfare chapter were incredibly useful to my project, and her own wonderful work on biometrics remains a rich source of scholarly inspiration. Janice Hladki encouraged me from my early days at McMaster, wrote me countless letters of support, and inspired my undergraduate thesis, master's thesis, and book topics. She continues to be a wonderful theorist, brilliant scholar, and fabulous pal. Mary Bryson heard my very first conference presentation and provided much-needed feedback and scholarly support. Michelle White's kindness to me as a young scholar at a particularly demoralizing time in my career was invaluable. Simone Browne's brilliant and innovative work on Canada's permanent resident card first made me start thinking about writing about biometrics at the border, a topic that became a central focus of this book. Big shout-outs to Barbara Crow and Sheryl Hamilton, who provided scholarly insights as I presented various pieces of my project. Mark Salter kindly encouraged me when I was still a graduate student. Jill Fisher, Caroline Andrew, and Diana Majury provided the mentoring a young scholar dreams of as I was writing this book. Simon Cole was incredibly generous in lending me his famous reputation to help me get my book published; *Suspect Identities* remains one of my favorite books. At Concordia University in Montréal Leslie Shade's wonderful work and encouragement were crucial as I wrote this book. Both Liz Miller's and Rae Staseson's feminist video interventions have inspired my own artistic work that is the complement to this book. Tagny Duff's innovative work in feminist bioart helped to push my own thinking about science and art combinations. A special mention must go to Kim Sawchuk and Yasmin Jiwani. I have been reading Kim's work with admiration since I began graduate school; her groundbreaking research at the intersection of feminist science studies and feminist media studies has continually inspired my own. She is also an incredibly generous mentor and colleague, and I hope to emulate her in both theory and practice, as she is that rare thinker who puts the two together beautifully. Yasmin Jiwani's wonderful scholarship and brilliant book continue to shape my entire theoretical framework. Like Kim, she is a true inspiration for how to practice theory and theorize practice. A research trip to UC Santa Cruz to speak with Donna Haraway about my work was an incredible experience and gave me a key intellectual lift at exactly the right time. Torin Monahan is an incredible and groundbreaking feminist thinker in surveillance studies. He also gave me innumerable brilliant suggestions on this manuscript that helped to make this a better book, and I owe a huge scholarly debt to him.

During my time as a postdoc at McGill I was incredibly fortunate to be supervised by Darin Barney. That year at McGill being supervised by him was one of the most fun experiences I've had in the academy. Getting to speak with him about his wonderful work on technology and communication was incredibly rewarding. He is one of the only people to have read every draft of every chapter of this book and to have provided brilliant suggestions on every single page. He has also modeled how to be a feminist academic who mentors with intellectual generosity and kindness, and I am indebted to him for years of both scholarly and professional advice. Thanks to Becky Lentz and Tamara Vukov for fun research chats. Enough good things cannot be said about Jonathan Sterne and Carrie Rentschler, who also provided me with mentoring and intellectual companionship both while I was in graduate school and then during my time hanging out with them while I was in Montréal. Their amazing work (and Jon's suggestion to further theorize failure) made my year there fun as well as academically stimulating. (Auditing Carrie's wonderful class in feminist media studies gave me innumerable ideas for how to improve my own thinking). Although I was only in Montréal for a short time, they reached out to me and brought me into their scholarly circle, which made McGill such a fabulous place to be that it was hard to leave. Jonathan also introduced me to his graduate student Tara Rodgers. It is impossible to capture how amazing it was working with her during my time at McGill and after, talking feminist science studies among other things. It always felt so weird that she was my research assistant, when often her advice was so insightful I felt I should be working for her. She was the one who advised me how to get my book published with Duke when she was still a graduate student, as her own edited book was already under contract!

At the University of Ottawa innumerable thanks must go to Ian Kerr and Val Steeves. Their support for my work was simply outstanding, and they continue to inspire with both their scholarly work and their incredible kindness and generosity to junior scholars. Ian Kerr was a generous mentor over the years that I was finishing up my PhD in Ottawa. Before I had a job or any intellectual community there he extended the incredible resources of the ID project, giving me office space, mentoring, and an intellectual community. Without his help it is doubtful I'd be where I am today. Val Steeves also continues to serve as a source of intellectual inspiration and was incredibly warm and generous in welcoming me and helping to find space for me at the University of Ottawa. My colleagues at the Institute of Women's Studies have provided me with a fabulous scholarly home. Thanks especially to Mélanie Knight during her time at the Institute, Dominique Bourque, Caroline Caron, Dominique Masson, Christabelle Sethna, Claire Turenne-Sjolander, and Denise Spitzer for wonderful mentoring and advice. Special thanks to our wonderful staff Margot Charbonneau and Michèle Phillips for incredible administrative support—they make it a joy to come in to the office. Thanks also to my colleagues elsewhere at the university, including Stephen Brown and Paul Saurette, for fun times and smart ideas. Finally, thanks to our graduate students, who continue to both push and nourish my research. They make coming in to work so much more rewarding and fun. A big thanks to Jessica Azevedo, Alexandre Baril, Ashley Bickerton, Caitlin Campisi, Sasha Cocarla (especially for her thoughts on theorizing kindness), Kinda Dalan, Samantha Feder, Emily Fortier, Heather Hillsburg, Sarah Lawrence, Virginie Mesana, Victoria Sands, Taiva Tegler, Ariel Trotster, Amanda Watson, Tanya Watson, Patrycja Wawryka, Amanda Whitten and Tori Whyte for all their brilliant insights. A particular thanks to Corinne Mason, who is a fabulous research assistant. It has been a true pleasure and incredibly stimulating to work with her on this and other projects.

Thanks to Ken Wissoker at Duke University Press for an incredibly kind and uplifting initial email indicating Duke's interest in the book. Working with my editor, Courtney Berger, has been a complete and total pleasure. Her wonderful advice, incisive comments, and smart thinking about how to improve this book have made it a million times better. I must also thank two anonymous reviewers for comments that improved this manuscript immeasurably.

As for my family, huge thanks to my mother, Sanda Rodgers, and her partner, Sheila McIntyre. Their encouragement and support sustained me, from listening to my terrible anxiety at ever finding a dissertation topic to offering gentle suggestions and encouragement at every step of the process. Should I be even half as fine a thinker, scholar, and feminist activist as they are, I would be satisfied. My mother talked tirelessly with me about the possibilities of my project. Her love and support during our morning talks were often what helped get me out of bed to face a long hard day of writing. It is impossible to thank her enough, though I have given it a try.

A huge thanks too to my friends and found family. Sarah Berry, Suzanne Bouclin, Robin Desmeules, Shaindl Diamond, Kristina Donato, Jackie Kennelly, Dana Krechowicz, Jamie Liew, Jena McGill, Cynthia Misener, Judith Nicholson, Sandra Robinson, Rakhi Ruparelia, Lilian Tu, and Emily Turk all sustained me during different parts of the process. Lisa Minuk's friendship and comedic genius have been a wonderful and rich addition to my life for the past ten years. There are many times when I don't know what I'd do without her. Sammi King's brilliance, but more her general awesomeness, have made life immeasurably better. Michael Orsini began as my colleague at the University of Ottawa but quickly turned into my new fave pal. His unbelievable depth and rigor as a scholar and his ear for sometimes daily phone calls during stressful times have much improved life. My best friends from my early undergraduate days, Shanta Varma and Helen Kang, continue to spend hours and hours listening to my melodrama every step of the way, patiently picking up the phone at 4 a.m. to calm me down. Helen's scholarly work within feminist science studies serves to nourish my own. Who else could I speak with for three hours a day about who knows what, day after day? Ummni Khan and Kathryn Trevenen have transformed Ottawa for me. Having them to talk through every facet of the everyday has changed my life. Ummni is a wonderful listener and just fabulous all-around person, not to mention supersmart thinker. I simply don't know what I'd do without her. I almost need a separate paragraph for Kathryn Trevenen. The year I was finishing the book was a particularly trying one, and she continually scraped me off the floor, patted me back into shape, reminded me that life was a fun place, generously gave me her seemingly endless supply of brilliant ideas, both scholarly and non, and generally made life a more beautiful place during difficult times.

Finally, huge, gigantic, tremendous thanks must go to my truly wonderful partner, Robert Smith? Robert offered everything from keen insights into my topic to explanations of what it means to have a Laplacian transformation to an attentive ear allowing me to rehearse every idea I ever had about this book to general hilarity in the day to day. Our life together is an enormous source of joy. I must have had a lucky star over my head the day I met him.

It is to my mother and my partner I dedicate this book.

introduction IMAGINING BIOMETRIC SECURITY

Mistakes are not abstract events or numbers, but lived phenomena everyone experiences and must reckon with.
MARIANNE A. PAGET, *THE UNITY OF MISTAKES*

The billboard advertising a new science fiction miniseries towers above passersby in Manhattan. Although it is unremarkable at first glance, those who stop to look at the ad unknowingly relinquish significant personal information. "Billboards that look back" are the latest in high-tech advertising. Each contains a tiny biometric camera that analyzes the viewer's facial features for gender and age (Clifford 2008). These "smart" billboards recall *Minority Report* (2002), Steven Spielberg's film about a society in which everyone is continually biometrically identified. The billboards are produced by the biometric company Quividi, which asserts that "the days of blind advertising are over." Quividi designed the billboards using technologies originally created for the Israeli military and adapted for commercial purposes. These fulfill the industry's desire for technologies able to reach specific markets rather than military targets. Using traditional gender markers in its high-tech marketing technologies, Quividi's smart billboards identify men with blue circles and women with pink circles (see their website). Although these biometric billboard cameras do not yet collect information on the viewer's race, Quividi asserts that the technology to do so is available and that race identification may soon be added to the collectible parameters (personal correspondence 2008). The color identification choices and symbols for race have not yet been determined. Companies marketing these technologies

FIGURE 1 The tiny biometric camera is located just below the TV screen on the left side of this advertisement for the science fiction miniseries *The Andromeda Strain*. In a striking point of similarity, the image of the man on the television is a doppelganger for Tom Cruise in *Minority Report* (Clifford 2008). Photo by Hiroko Masuike for the *New York Times*. Courtesy of Redux Pictures.

promise that the collection of this information will allow companies who purchase it to refine their promotional practices, targeting "one advertisement to a middle-aged white woman, for example, and a different one to a teenage Asian boy" (Clifford 2008).

The example of "billboards that look back" touches upon a number of the central themes of this book. *When Biometrics Fail* engages questions raised by the role of state institutions and the military in driving technological development and expansion, the relationship between surveillance and marketing, the permeable boundary between science and popular culture, and our desire to read identity off the body. What does this book tell us that is new about our lives? Human bodies are not biometrifiable. Despite the multibillion-dollar investments in these technologies, investments that depend upon the assumption that bodies can be easily rendered into biometric code, they cannot. Biometric science presupposes the human body to be a stable, unchanging repository of personal information from which we can collect data about identity. Biometric failures, encompassing mechanical failure, failures to meet basic standards of objectivity and neutrality in their application, and the failure to adequately conceive of the human subjects and identities that are their purported objects, necessarily call these claims into question. Yet despite persistent mechanical failures, biometric technologies still accomplish a great deal for state and commercial actors whose interests are tied to contemporary cultures of security and fear. In this sense biometric technologies succeed even when they fail. On the other hand, even when they function technically, biometrics do real damage to vulnerable people and groups, to the fabric of democracy, and to the possibility of a better understanding of the bodies and identities these technologies are supposedly intended to protect. In this sense biometric technologies fail even when they succeed. The case studies explored in this book cast doubt on scientific and industry assertions that the human body can be made to speak the truth of its identity through the use of biometric technologies. In examining those instances when biometrics break down, we see that the real-world deployment of biometric technologies depends upon practices of inscription, reading, and interpretation that are assumed to be transparent and self-evident and yet remain complex, ambiguous, and, as a result, inherently problematic.

This book takes biometric failure as its starting point of analysis. Arguing for the application of the language of failure not only to those technologies that do not function reliably, I adopt failure as a narrative framework through which to think about how biometric technologies are implicated in what feminist theorists refer to as "interlocking systems of oppression" or a "matrix of domination" (Fellows and Razack 1998; Collins 1990). Historical accounts of technological failure often present it as self-evident in an "I know it when I see it" approach to technical breakdown. This type of narrative usually goes something like this: a technology is developed, but due to persistent mechanical problems or the lack of a consumer market, it fails to work properly or sell consistently, resulting in its abandonment. Successful technologies are represented as natural technological outgrowths of earlier, unsuccessful models, producing stories of technical failure as the productive prototypes of more profitable and reliable technologies. In this narrative, which is commonplace (Petroski 1994; Adams 1991; Levy and Salvadori 1994; Casey 1998), failure is a "cautionary tale" (Lipartito 2003) leading to the development of better-functioning technologies. And yet the suggestion that technological failures are obvious is deeply problematic (Gooday 1998; H.-J. Braun 1992; Lipartito 2003). For example, how do we codify the temporality or spatiality of failure? That is, how often does a technology have to break down before it is considered unreliable? How long does it have to sit on the shelf before it is considered a commercial failure? What if a technological artifact enjoys success in the basements of underground technophiles but not in broader commercial spaces? Just as technical successes both produce and are produced by existing cultural frameworks and ideologies, so are technical failures (Lipartito 2003), or in the words of the historian Graeme Gooday (1998), technological failures demonstrate a high degree of "interpretive flexibility."[1]

A theoretical point that runs through this book is what Donna Haraway (1997:142) terms "corporeal fetishism." That is, how does contemporary biometric discourse produce maps of complex living bodies that render them autonomous things-in-themselves rather than actors in networks of interrelationships? One example of an offshoot of corporeal fetishism (or "genetic fetishism") is Richard Dawkins's (2006:156) study of the "selfish gene" as a scientific object that may be examined in isolation from its biological and cultural context. Here genes are described as discrete, autonomous agents containing the "blueprints" for life. Moreover genetics as a science is imagined to be outside troping or representational strategies. Haraway describes the representation of genes and bodies as independently functioning things as a form of fetishism that produces "interesting 'mistakes'" in which we confuse a complicated network of lively relating for a fixed thing (what Haraway refers to as "life itself"). That is, the suggestion that a gene holds the blueprint for life is a form of "mistake" where the map or representation of a thing is confused with "the doings of power-differentiated lively beings" (1997:135). These mistakes produce particular kinds of subjects and objects (136).[2] As a result corporeal fetishism produces a problematic kind of cartography in which the model becomes a fetish object such that it is no longer understood as a form of representation but instead as a technoscientific truth. The situatedness of knowledge is forgotten, including the ways that these maps draw on existing cultural assumptions about gender, race, sexuality, disability, and class.

Thus, the assertion by biometric scientists and biometric industry officials that binary maps of bodies produce bodily truths is a form of corporeal fetishism by which bodies are transformed into reified "things," objects imagined to exist outside culture. In this book I investigate how biometric renderings of the body come to be understood not as "tropic" or "historically specific" bodily representations, but instead are presented as plumbing individual depths in order to extract their core identity. This is a primary way that biometric technologies fail. Biometric industry assertions that identities are no more than their binary code is a "philosophical-cognitive" error mistaking a mapping or troping narrative for a "concrete entit[y]" (Haraway 1997: 147). This type of error has significant implications for the subjects and objects of biometric technologies. Bodies rendered biometric become a particular kind of capital. For example, biometric maps of the body help to spin the bodies of prisoners, welfare recipients, and travelers into valuable data. Biometric representations of the body also produce new forms of identity, including unbiometrifiable bodies that cannot be recognized by these new identification technologies, a subject identity that has profound implications for individuals' ability to work, to collect benefits, and to travel across borders.

Biometric failures are wide-ranging. They occur across biometric technologies and consistently plague the introduction and application of these technologies to any institutional setting or marketing program. These new identification technologies suffer from "demographic failures," in which they reliably fail to identify particular segments of the population. That is, even though they are sold as able to target markets and sell products to people specifically identified on the basis of their gender and race identities, instead these technologies regularly overtarget, fail to identify, and exclude particular communities. For example, biometric fingerprint scanners are consistently reported to have difficulty scanning the hands of Asian women (Sturgeon 2004; Nanavati, Thieme, and Nanavati 2002), a category not problematized in the scientific literature. Iris scanners exclude wheelchair users and those with visual impairments (Gomm 2005; Nanavati, Thieme, and Nanavati 2002). More generally, "worn down or sticky fingertips for fingerprints, medicine intake in iris identification (atropine), hoarseness in voice recognition, or a broken arm for signature" all give rise to temporary biometric failures (Bioidentification 2007). More durable failures include "cataracts, which makes retina identification impossible or rare skin diseases, which permanently destroy a fingerprint" (Bioidentification 2007). This broad range of biometric failures to identify some subjects and to overselect others has given rise to questions by the media and privacy advocates as to whom exactly biometric technologies reliably can identify.

One form of biometric failure that has generated significant anxiety is the ability of the scanner to recognize whether a body is alive or dead. Biometric breakdown inhabits even this most basic of differences. Science fiction, a genre that often reveals preoccupations about technologies not yet found in public debates, has dramatized this type of "identity theft." Fictional commentaries like the science fiction series *Dark Angel* (2000) suggested that people might have their body parts removed and then used to mimic biometric identity in order to access secure points of entry.[3] *Minority Report* was one of the first films to showcase the potential of biometric technologies for both surveillance and marketing. As such this film is a key site for the production of biometric meanings. In a particularly gruesome scene, the central character has his eyes removed and replaced with those of another man in order to hide his identity from iris-scanning technologies.

Biometric companies dismissed these scenarios as science-fiction fantasies, assuring the public that "liveness," or proof that the identity came from a live body, was needed for the establishment of biometric identity. Their assertions were specifically challenged when the S-class Mercedes (equipped with a biometric fingerprint scanner on the ignition) belonging to a Malaysian businessman was stolen. The thieves forced the owner of the car to "put his finger on the security panel to start the vehicle, bundled him into the back and drove off" (Kent 2005). Feeling frustrated at having to get the owner out of the trunk of the car each time they wanted to start the engine, the thieves cut off his finger with a machete and dumped him at the side of the road. They then used the severed finger to start the car without difficulty, fleeing with both car and finger. Reports of people having their fingerprints surgically removed to avoid identification by biometric fingerprint scanners also began to make news headlines ("Criminals" 2008). These accounts suggested that scenarios formerly relegated to the realm of science fiction might soon be coming to a theater of the real near you.[4]

Although biometric failures are wide-ranging, knowledge of them is limited. The reasons for this are complicated. Biometric failures are often treated as aberrations, as exceptions, or as caused by a few incompetent scientists who did not fully refine their technological products before releasing them onto the market. And yet in examining the application of these new identification technologies to a broad range of institutions and programs, I find that rather than a small minority, biometric errors are endemic to their technological functioning. In her book *The Unity of*

Mistakes Marianne Paget (1993), a sociologist of medicine, examines the complex meanings of medical errors. She finds that although the failures of medicine are regularly presented as temporary setbacks caused by a small number of incompetent ("bad apple") physicians, in fact mistakes are part of the practice of medicine (4). The theme of a "few bad apples" is common in the study of failure. In Sherene Razack's (2004) examination of the failure of Canadian peacekeepers to safeguard human rights during their mission in Somalia, instead themselves participating in the torture and murder of Somali civilians, she argues that the impulse of Canadian national inquiries into these atrocities refused to acknowledge them as endemic. Instead official inquiries into the torture of Somalis by Canadian peacekeepers attempted to dismiss them as exceptional, the result of a few high-level officials gone bad, or a "few bad apples" (117), rather than the result of systemic and widespread racist beliefs within Canadian peacekeeping forces. Razack points out that to look for exceptional failures on the part of Canadian peacekeepers was to "miss the absolute ordinariness and pervasiveness of racist attitudes and practices both in the military generally and among troops deployed to Somalia" (117). I investigate biometric failures as ordinary, as both endemic and an epidemic, rather than as unusual or the result of a few flawed scientists.

We have limited ways of speaking about scientific errors. In studying the reasons for the silence around medical errors, Paget found that the denial of failure and the use of technical terms to mask the subjectivity of medical errors came in part from the norms of masculine, scientific culture as well as from individual longings for a high-stakes science that would get it right every time.[5] In addition a relentless quest for "mechanical objectivity" in which science is imagined to exist outside culture—including outside cultures of racism, sexism, and economic inequality—provides part of the motivation to deny failure related to systemic forms of discrimination. We continue to see the perpetuation of scientific norms that uphold a scientific practice free from cultural assumptions about identity as a vaunted scientific goal (Daston and Galison 1992; Gould 1996). Developed in a culture deeply divided by interlocking oppressions and with the aim of being able to reliably identify othered bodies, biometrics fail to work on particular communities in ways connected to race, class, gender, sexuality, and disability. I argue that we need to think beyond the "few bad apples" theory of failure and think about structural failures related to systemic inequalities more broadly.

It is useful to review the nature and range of biometric technologies. Biometrics is the application of modern statistical techniques to measure the human body and is defined as the science of using biological information for the purposes of identification. Biometric technologies include, but are not limited to, digital fingerprinting, iris scanning, and facial recognition technologies. Biometrics take an electronic reading of the body, an identification process that communication scholars note is at the intersection of multiple communications and information processes, including photography, videography, computer networking, pattern recognition, and digitalization (Gates 2004). Although analog forms of biometric science date back to the nineteenth century, the focus of this book is on digital biometric technologies. These are distinguished by their ability to store bodily information in binary code and to share this information easily within networks. Biometric technologies are the latest development in a long line of technologies that measure the body for the purpose of identification, ranging from anthropometry to fingerprinting. Breaking the body down into its component parts, from retina to fingerprint, biometric technologies purport to make individual bodies endlessly replicable, segmentable, and transmissible in the transnational spaces of global capital.

Biometric images of the body are found at the center of securitization, industrialization, commodification, and the body in the age of mechanical reproduction, what the communication theorist Jonathan Sterne (2003) calls "a thoroughly modern moment." These new identification technologies increasingly pervade contemporary life. Personal laptops contain biometric fingerprint scanners, fitness centers capture their clients' biometric information to enable patrons to leave their wallet at home, cars are outfitted with biometric locks, and biometric facial recognition technology is now included in all U.S. and Canadian passports. The biometric industry, scientists, and institutional investors all regularly represent biometric technologies as a fail-safe solution to all of our problems, whether marketing, crime, or security. News articles, scientific studies, and policy documents such as the *9/11 Commission Report* all proclaim that biometric technologies are crucial to preventing the next terrorist attack. We are told that the stakes are high and that we forgo biometrics at our own risk. Without biometrics, crime will rise unchecked and terrorists will flood across U.S. borders. Biometric technologies are necessary to protect us from a dangerous world, one in which we need increased monitoring and surveillance to keep us safe and where, if we have nothing to hide, we have nothing to fear.

When Biometrics Fail calls these claims into question. I examine our growing surveillance society by analyzing three case studies: the adoption of biometric technologies by the prison industrial complex, the welfare system, and the Canada-U.S. border. In *Surveillance in the Time of Insecurity* Torin Monahan (2010) argues that discourses emphasizing external threats to the state (national security) are trumping threats to vulnerable groups within the state (human insecurity). Following Monahan, I argue that what is at stake with respect to biometrics is quite different from what mainstream discourse about these technologies would have us believe. Contesting industry assertions that the primary impact of biometrics is to improve national security, I show that biometric science codifies existing forms of discrimination and that biometric technologies raise troubling implications for both substantive equality and democracy. I analyze biometric technologies as a problem, in the richest sense of that word. In my study of the origins of these new identification technologies I show that state institutions deploy biometrics to enact institutionalized forms of state power upon vulnerable populations. Rather than providing a straightforward answer to what ails us, biometric technologies raise a Klein bottle of questions.

When Biometrics Fail reviews the history and development of biometric identification technologies with particular attention to their expansion from law enforcement to welfare and national security. Biometric technologies have a long history of surveillance applications. They first were designed in the 1960s for law enforcement and military applications. In the 1990s biometrics were adopted for use in the surveillance of welfare recipients. It was in the post-9/11 era that biometric technologies really came into their own, and industry profits began to soar (Feder 2001) alongside the expansion of a security industrial complex. Identified by the *9/11 Commission Report* (National Commission on Terrorist Attacks upon the United States 2004) as the key technology in which the state should invest, biometric technologies were expanded to the borders of the United States, including the formerly unfortified U.S.-Canada border. Relying on archival materials consisting of scientific reports, government documents, media coverage, industry publications, and legal records, I map the growth and expansion of these technologies in the prison system and the welfare system and at the border. I examine biometric adoption as a crucial point at which institutions, industry, and individuals wage contests for the meaning of these new identification technologies. In looking at these three "crisis points" of the real-world deployment of biometrics, I reveal the interests and institutions driving their growth beneath the invisible fabric of everyday life.

John Peters (1999) argues that we constantly construct communication as a perfectible art. If only we had a better communication technology and an ideal set of circumstances, communicative exchange could be flawless. And yet, as Peters reminds us, a "mistake is to think that communications will solve the problem of communication, that better wiring will eliminate the ghosts" (9), when communication itself is the story of miscommunications, of failed exchanges, and of unexpressed longings for understanding. The complex communicative acts between biometric scanners, individuals, and institutions are, like any communicative exchange, filled with error. Claims about the possibilities offered by biometric identification are everywhere. We are told that these technologies can read the body perfectly, from the tips of our fingers to the depth of our thoughts, providing reliable proof that we are who we say we are. The biometric industry calls for greater use of security technologies to keep us safe, while these same ideologies of "more technology, more security" are used to bolster and sell biometrics. Few have stopped to ask "What will this cost?," "How does this work?," or even "Does this work?" These are the questions at the heart of *When Biometrics Fail.* In asking what happens when biometric technologies fail and what those failures tell us as well as through an investigation of the cultural and historical processes that gave rise to biometrics, I seek to understand the significant political, economic, and social consequences involved in the broad deployment of these technologies. In doing so I continually rethink assertions that are regularly given as unproblematic truths, including what it means to say that borders are obvious, security is natural, and new technologies work perfectly. The U.S. government's latest biometric initiative, the world's largest biometric database, is currently scheduled to cost $1 billion (Hruska 2007). The Newborn Screening Saves Lives Act of 2007 now allows the U.S. government to collect biometric information at birth from every baby born on American soil (Aldridge 2008). This is an opportune and necessary moment to examine the origins and expansion of biometric technologies.

The biometric industry represents these technologies as particularly useful because they are to replace human subjectivity with "mechanical objectivity" (Daston and Galison 1992:98), eliminating human handprints from scientific practice. Chief among mechanical objectivity's virtues is its imagined ability to labor continuously without the danger of succumbing to human error or subjective judgment (83). The industry claims that biometrics are able to transfer subjective human decision-making processes to objective machines, rendering human assumptions invisible. A CNN article on the future of airport security assures us that the new biometrics will be able to use "brain-fingerprinting technologies" (rather than race, class, or religion) to predict dangerous behavior (Rosenblatt 2008). Smart biometric rugs and other new biometric technologies will detect threatening conduct by first presenting stimuli and then measuring signs of physical distress such as heart rate, body temperature, and respiration. Scientists and institutional adopters of these new identification technologies describe them as free from human assumptions about identity: "Although traditional security profiling can discriminate by race and religion, security experts say behavioral profiling is more fair, more effective and less expensive" (Rosenblatt 2008). Yet the stimuli are far from neutral. Rosenblatt reports that one suggestion is the subliminal repetition of the words "Osama bin Laden" or "Islamic jihad" written in Arabic. Far from being identity-neutral, biometric technologies can be designed specifically to profile othered bodies, including those of Arab and Muslim communities.

Biometric forms of corporeal fetishism are productive. Beyond the simple translation of a material body into binary code, biometric maps of individuals produce new understandings of the human body. Secretary Michael Chertoff of the Department of Homeland Security argued, "A fingerprint is hardly personal data because you leave it on glasses and silverware and articles all over the world" (Aldridge 2008). Suggesting that human beings continually leak biometric information into public space, Chertoff uses biometric discourse to reimagine the body as public. This has dramatic ramifications for how we understand privacy and personal information. Biometric discourse thus produces individual bodies as publicly available human inventory. In the NEXUS frequent-flyer program subscribers store their iris information in a database in order to speed their passage across the U.S.-Canada border. By enrolling in the program, NEXUS travelers gain the privilege of bypassing customs agents and going directly to a scanner for identity confirmation. The directions are:

- Proceed to the self-serve kiosk located in the Canadian inspection services area.
- Stand in front of the camera to have your irises captured. Follow the prompts on the screen and complete the entry process. You will then receive a self-serve kiosk receipt.
- The self-serve kiosk will direct you to the cashier for collection of duties and taxes owing, if applicable. (Canadian Border Services Agency 2009)

Allowing travelers to scan their body and providing them with a receipt of the transaction produces new understandings of the "body as commodity." Biometrics break bodies down into their component parts in ways that allow them to be marketed more easily in the transnational marketplace, whether as a security risk or a potential consumer (S. Browne 2009). Unlike the unruly material body, biometric bodies offer up a text over which scientists may have absolute control. In this way biometric technologies make the body into a thing, a more easily governable entity, in a process of corporeal fetishism. The flimsy material body is rendered rugged as biometric technologies make the body replicable, transmittable, and segmentable. The knowledge generated by the use of biometrics to test identity is asked to perform the cultural work of stabilizing identity—conspiring in the myth that bodies are merely containers for unique identifying information which may be seamlessly extracted and then placed into a digital database for safekeeping. Key to this system is individual compliance with this new imaging process, allowing biometric science to scientifically manage the body more efficiently, and seemingly without error.

But biometric discourse produces more than new understandings of the human body. Rising to prominence at a time when the state is determined to make citizens newly visible for the purposes of governance, biometric technologies are deployed to produce new understandings of security, the border, and the nation-state. Biometric errors are also productive: they produce some bodies as belonging to the nation-state while excluding others. In doing so biometric technologies inform new understandings of difference. Thus even when biometrics fail to work, they are successful in other ways. For example, before the introduction of biometrics, passports contained "digitized photos, embossed seals, watermarks, ultraviolet and fluorescent light verification features, security laminations, microprinting and holograms" (U.S. Department of Homeland Security, Bureau of Customs and Border Inspection 2006:16). The addition of biometric technologies is used less to make identification possible than to convince consumers that they are now safer, whether or not security is actually improved.

When Biometrics Fail is not simply a story of the inability of biometrics to function seamlessly, with the assumption that these technologies eventually will be refined and improved in laboratories sterile to human culture. In investigating their multiple failures I denaturalize the development of biometric technologies as the triumph of a new digital identification technology over its outdated analog ancestors. Calling into question assertions by biometric scientists and the biometric industry that these new identification technologies are simply more accurate and efficient than old-fashioned forms of identification such as documents and photographs, I refute the assumption that it is possible to develop a perfect security technology able to definitively map the body, solve complex social problems, and keep the nation safe. Instead I raise broad questions about the meanings of biometric failure. Thinking about the productive nature of error, I investigate mechanical malfunctioning as well as how biometrics fail to work in the ways that industry claims they do, that is, objectively and neutrally.

My analysis of biometric failures provides an opportunity to think about the impoverished language we have for speaking about scientific error and ambiguity. The scientific literature on biometric technologies reveals a long list of common malfunctions. From "failure to enroll" and "failure to capture" rates to "false accept" and "false reject" rates, many kinds of technical malfunctions are documented. And yet these discussions of biometric errors are limited to technological crashes. A central aim of this book is to broaden existing conversations about the pros and cons of biometrics from discussions of efficiency, privacy, and security to consider the relationship of these technologies to inequality, whether it is the ability to function reliably and efficiently, to save the state money, or to operate free from systemic forms of discrimination. Whether facilitating the surveillance of poor folks through the exchange of biometric information between welfare and law enforcement agencies or collecting biometric information from marginalized bodies in order to better police the edges of the nation-state, the adoption of biometric technologies by state institutions raises troubling implications for racism, sexism, ableism, and classism as well as homophobia and transphobia.[6]

In a climate in which some believe technologies will provide magic-bullet solutions to complex social problems, biometric discourse fails to acknowledge the ways that biometric technologies code their cultural context. Biometric discourse thus makes the economic, political, and social stakes of adding new technologies to human space invisible. The ways that citizens are mobilized around biometric technologies—as well as the ways that these technologies are mobilized by the biometric industry, the mainstream media, and state institutions interested in adopting them—limit the ways that we conceive of their implications. Biometrics are not about simple improvements to security. At stake in the application of biometric technologies to state programs are constrained possibilities for substantive equality, democracy, and socially just forms of subjectivity. At the heart of *When Biometrics Fail* is the paradox that biometric technologies are deployed in the name of freedom at the same time as they hold particular bodies static through the production of new forms of imprisonment and immobility.

Borrowing from both Foucauldian and feminist science studies methodologies, I examine the origins of biometrics and demonstrate why this is crucial to understanding and assessing their current expansion. The case studies each make clear that questions of race, gender, disability, and sexuality remain central concerns of and about the use of biometrics. Although the discourse of biometrics claims to render the body free from markers of identity, questions about identity remain central to the concerns that biometrics are deployed to address.

In chapter 1, "Biometric Failure," I define biometrics through a detailed study of the science behind these new identification technologies. A primary aim of biometric science is to produce technologies capable of rapid identification given the large numbers of people who pass through biometric scanners daily. One way to speed biometric identification is to reduce the size of the database against which a particular individual is checked. Biometric scientists have a number of ways of doing this, one of which is by relying on "soft biometrics." Soft biometrics explicitly include race, dividing participants into racial categories based on facial measurements. Other strategies organize the data according to gender, categorizing persons based on whether they have short or long hair or whether they are wearing masculine or feminine clothing. In examining the science that constructs biometrics, I demonstrate that assumptions about gender, race, sexuality, and disability are encoded by scientists directly into the operational elements of the technologies, revealing the inseparability of science from culture. I conclude the chapter by examining mechanical biometric failures and their connection to bodily identities.

Chapter 2, "I-Tech and the Early Beginnings of Biometrics," locates the development of biometric technologies in their origins in the law enforcement and prison systems in the 1960s and 1970s. I first give a brief history of the development of each biometric technology, from biometric fingerprinting techniques to retinal and iris scanning. I show that biometric technologies draw on a nineteenth-century desire to force the body to speak the truth of its identity, much like Bertillon's early system of identification, which aimed to create a cage of information from which the criminalized body could not escape. I situate these technologies against the backdrop of the expansion and increased profitability of the prison system, and document the part that biometrics play in a return to earlier attempts to read criminality off the body. I describe the consequences of replacing human labor with biometric technologies in the prison system, including the dangers of using prisoners as a test bed for new technologies, and consider the ramifications of thinking about prisoners as human inventory and the possibilities that biometric technologies produce for their increased surveillance.

In chapter 3, "Criminalizing Poverty: Adding Biometrics to Welfare," I examine the first major expansion of biometric technologies in the 1990s from the prison to the welfare system. Marketed as a cost-saving technology that would reduce welfare fraud, the adoption of biometrics was a politically popular move following Reagan's welfare reform strategies of the 1980s. This initiative proved to be extremely profitable for an industry with no prior obvious markets. Although welfare agencies justified the adoption of biometric technologies as cost-saving measures, no cost-benefit analyses were conducted prior to their introduction into any state in the U.S. In California, the first state to adopt biometrics for the welfare system, auditors found that the technology cost the state millions more than it saved. When the technologies were expanded to states beyond California, they often proved so expensive that they subsequently were eliminated due to insufficient state funds to pay for them.

At the same time the intensification of the criminalization of poverty complicated the discourse on welfare fraud. In contextualizing the expansion of biometrics, I consider why biometric technologies designed for the prison system were next used in the administration of benefits for individuals receiving state aid. I demonstrate that biometrics were part of a larger move by the state to use quick-fix technologies as the solution to complex social problems, including the criminalization of poverty. Adding biometric technologies to welfare produced a number of technological externalities. These included the creation of new categories of disability, the sharing of information between welfare and law enforcement, and the tracking of survivors of domestic violence. Their introduction also disproportionately affected immigrants and refugees receiving state aid.

In chapter 4, "Biometrics at the Border," I investigate the rise of biometric technologies to center stage as part of the post-9/11 security complex at the U.S.-Canada border. Before 9/11 the border was often described as the "longest undefended border in the world." This narrative of "Canadian exceptionalism" imagined Canadians as white, a representation with concrete privileges for Canadians at the northern U.S. boundary until September 11, 2001. Following the attacks on the Pentagon and the World Trade Center, and in particular as a result of false press reports that the 9/11 terrorists had entered the United States from Canada, the U.S. ceased to regard Canada as its friendly neighbor to the north. Instead U.S. media depicted Canadians as potential terrorist threats, an understanding largely attributed to Canada's more generous immigration and refugee policies. As Canadian bodies underwent racialization at the U.S. border, a technology able to definitively visualize newly inscrutable Canadian bodies was understood to be essential, and biometric technologies were identified as able to accomplish this task. Featured in all U.S.-Canada border accords signed after 9/11, biometric technologies were central to the transformation of Canadians in the U.S. cultural imaginary. This dramatic shift in the representation of the northern border of the U.S. provides a window into state-making in the age of security. In examining the role of biometrics at the edges of the nation-state, I provide "before" and "after" snapshots of these new identification technologies as they are enlisted in the redefinition of the spatial and cultural landscape of the U.S.-Canada boundary. I also examine the increased calls to "outsource" border security, using biometrics as the technological strategy to do so. These new identification technologies are called upon to move the border away from the material edges of the U.S. state, particularly through the use of policies mandating the biometric identification of travelers before they leave their country of origin.

In chapter 5, "Representing Biometrics," I examine the depiction of biometrics as perfect identification technologies, in which they are imagined to be capable of engraving a unique identifier upon the scanned body that will reliably indicate that you really are you. From science fiction to news stories, these depictions of biometrics play a central role in educating the public about these technologies. In my analysis of a photograph and cartoon that were widely circulated on the Internet, each making a visual argument for the outdated nature of identification technologies like photographs and ink fingerprints, I examine the role of representation in the expansion of these technologies. I also consider the importance of the claim that biometric identification technologies do a better job than fingerprints and photos in their imagined ability to tell racialized bodies apart. Assertions that biometric technologies are no longer "the stuff of science fiction," that "biometrics aren't just for the *X-Men*" or that "they work just as well as they do on *Star Trek*" are commonplace in policy debates about these technologies, revealing that the seamless functioning of biometrics in popular media texts has a significant impact on their expanded use by the state. I also examine how biometric maps of the body produce new "practices of looking" (Sturken and Cartwright 2001:10); rather than "passive acts of consumption" (42), looking produces new ways for us to "express ourselves, to communicate, to experience pleasure, and to learn" (2). Biometric practices of looking include an emphasis on the identification and verification of individual identities, as well as what I term "surveillant scopophilia." Extending feminist analysis of scopophilia—meaning to take pleasure in looking—I argue that biometric maps of individuals produce new forms of pleasure in looking at the human body disassembled into its component parts while simultaneously working to assuage individual anxieties about safety and security through the promise of surveillance.

In the conclusion I review the ways that biometric technologies are regularly dehistoricized and held up as new, utopian technologies that promise to solve the persistent problem of identification. This claim is offered as justification for the dramatic expansion of their use. The conclusion sums up my intervention in this narrative of unqualified technological progress. I describe how biometrics arose out of a conversation among industry specialists, engineers, computer scientists, marketers, and those who willingly gave up or were compelled to give up their personal information. I conclude by examining the ways that biometrics are a "politically successful policy failure"[7]—in which success is defined primarily in terms of whether the technology *appears* to be keeping us safer rather than by marked or measurable improvements in our security—and at what price and to whom.

one BIOMETRIC FAILURE

Stop! Right now, think of how many passwords and personal identification number (PIN) codes you have to remember. How often do you forget them? It is very inconvenient to remember those codes. Now do you have your fingers, eyes, voice, and face with you? The answer hopefully is yes! Have you ever forgotten any of those body parts? Not very likely! What if you could use those body parts instead of passwords and PIN codes to verify who you are? Would that not be more convenient? It also seems logical that it could be a more secure way of authenticating a person.

PAUL REID, *BIOMETRICS FOR NETWORK SECURITY*

Biometrics are celebrated as perfect identification technologies. They will secure your laptop, identify terrorist threats, and reduce crime by stabilizing the mercurial identities of criminalized individuals. As in the epigraph that begins this chapter, biometric scientists imagine these technologies as more highly evolved, efficient, and accurate versions of older techniques of identification. Why have to remember a password or PIN code when your body can easily replace it—a body that can now be digitalized? Isn't it logical that digital technologies function better than their analog counterparts? Much is made of the potential benefits biometric technologies can bring to the knowledgeable consumer in this textbook, published in 2004:

> With the cost of biometric technology falling, it will soon be reliable and cheap enough for everyday use. For example:
>
> - Enter your favorite coffee shop in the morning, and your coffee order is waiting for you at the counter.

- Your daughter's boyfriend, whom you do not like, comes to the house and the door won't open to let him in.
- Your own little personal identification device recognizes someone you have not seen in years and provides you with his/her name. This could result in no more embarrassing blank looks on your face. (P. Reid 2004:231)

In these examples biometric technologies are able to supply simple solutions to domestic problems, from patriarchal protection to the limits of long-term memory. Not restricted to the high-stakes tasks of securing the border or ending crime, biometrics are able to additionally (and seamlessly) break down the quotidian into moments for market, as scientists imagine these technologies to work perfectly, identifying and verifying individual bodies without error while providing a new range of consumer services. And yet biometric technologies do not work in the straightforward manner such discourse would have us believe, whether industry representations, government accounts, or laws and policies mandating their deployment. Although biometrics are explicitly sold to us as able to circumvent problematic human assumptions about race, gender, class, and sexuality; in fact, this book demonstrates that it is upon rigid and essentialized understandings of race and gender that these technologies rely.

In this chapter I analyze the science constructing biometric technologies and show how they rely upon outdated and erroneous assumptions about the biological nature of identity. In investigating the scientific principles upon which biometrics are based, I ask whether these new identification technologies perpetuate existing forms of inequality. I start with the way the biometric industry describes how the technologies work, then examine the assertions about the alleged potential of these technologies. In contextualizing these claims I demonstrate that biometric technologies are the latest in a long line of identification technologies claimed to be impartial and objective. Considering the ways that culture is always encoded into technology, I revisit the central question of this book: What happens when biometrics fail? I show that although biometrics were marketed on the basis of their ability to avoid human pitfalls, these very same human assumptions about the biological nature of identity are encoded right into these new identification technologies.

Defining Biometrics

Biometrics is the science of using biological information for the purposes of identification or verification.[1] Although the term *biometrics* dates back to the early twentieth century and was used to refer to mathematical and statistical methods applied to data analysis in the biological sciences, the focus of this book is on digital biometrics. A biometric attribute is defined as a "physical or psychological trait that can be measured, recorded, and quantified" (P. Reid 2004:5). The process of acquiring the information about the physical or behavioral trait—whether a digital fingerprint, iris scan, or distinctive gait—and then storing that information digitally in a biometric system is called *enrollment* (P. Reid 2004:6; Nanavati, Thieme, and Nanavati 2002:17). A *template* is the digital description of a physical or psychological trait, usually containing a string of alphanumeric characters that expresses the attributes of the trait (P. Reid 2004:6). Before the biometric data are converted to a digital form, they are referred to as *raw data* (Nanavati, Thieme, and Nanavati 2002:17). Raw biometric data are not used to perform matches—they must first be translated into a biometric template before they can be utilized—a process that is achieved with the help of a biometric algorithm. *Biometric algorithms* are frequently described as recipes for turning a person's biological traits into a "digital representation in the form of a template" (P. Reid 2004:6). This recipe is usually proprietary, and it is what a biometric technology company sells, arguing that their recipe for creating a template from raw biometric data is better than another company's recipe.

Vendors represent biometric technologies as able to answer two questions. The first question refers to identification and asks, Who am I? Often described as a 1:N matching process, the presentation of a biometric template created in real time (called a *live biometric*) is checked against a database of stored biometric templates. Used more commonly in security and law enforcement applications, this process allows for one person to be checked against a list of persons (P. Reid 2004:14). The second question that biometric technologies are imagined to be able to answer concerns verification: Am I who I say I am? Referred to as a 1:1 matching process, verification checks the presentation of the live biometric with the person's template stored in the database to determine if they match. If the live biometric is the same as the stored biometric, there is a match and the identity is verified. Verification is held up in biometric discourse to be a much quicker process than identification, since it must check only one live biometric against one stored biometric, rather than checking a particular biometric against an entire database. Biometric technologies that rely on verification are more commonly used in physical and informational access applications, including secure building and computer network access (14).

The industry claims that a central innovation provided by the technology is that biometric information can neither be stolen from nor forgotten by the individual, as it depends on the measurement of physical or behavioral traits. As biometric technologies measure bodies and bodily behavior (such as gait), the "provider of the trait always has them with him or her" (P. Reid 2004:3) and cannot substitute another's information. Industry representatives thus claim that biometric technologies provide convenience, security, and accountability (Nanavati, Thieme, and Nanavati 2002:4–5).

The biometric industry does acknowledge that errors do occur and cites three errors as particularly common (P. Reid 2004; Nanavati, Thieme, and Nanavati 2002). One is the *false acceptance rate*, in which a person who is not you is accepted as you. The *false rejection rate* occurs when you are not accepted as you. This is usually given as a percentage of the chance of somebody else erroneously being identified as you. Another type of error is the *failure to enroll*, often given as a percentage of the possibility of someone failing to be enrolled in the system at all (P. Reid 2004:6). In *Biometrics for Network Security* Paul Reid asserts that current problems with biometric technologies "will be solved with time, money and technological advances" (121).

Biometric technologies are broken down into two categories: active biometrics and passive biometrics. *Active biometric technologies* depend on the user actively submitting information to a biometric scanner. For example, finger scanning relies on the person actively placing a finger on the plate in order to have the print captured. *Passive biometric technologies* allow for the covert collection of biometric data, as in the case of smart surveillance cameras able to secretly capture a person's facial biometrics. Biometric technologies currently in use include fingerprint imaging, hand geometry, iris scanning, voice biometrics, and facial recognition technology. Other biometric technologies in development for commercial use include gait and vein recognition, signature and keystroke technologies, and the use of DNA for the purposes of identification or verification, and the industry has endorsed a number of futuristic technologies, from olfactory recognition to recognition based on thought patterns (brain mapping).

Biometric Objectivity

As conversations on whether the state is racially profiling particular communities shift to whether racial profiling is an important and necessary state practice (Bahdi 2003), biometric discourse admits that racial profiling is occurring, but suggests that it can be overcome using new technologies. For example, media and scientific reports regularly depict biometrics as able to circumvent discrimination. One industry representative, Frances Zelazny, director of corporate communications for Visionics (a leading U.S. manufacturer of biometric systems), asserts that the corporation's newly patented iris-scanning technology "is neutral to race and color, as it is based on facial features recognized by the software" (Olsen 2002). Zelazny suggested that biometric technologies' impartiality helps to protect the privacy of individual citizens: "That the system is blind to race is a privacy enhancing measure" (Olsen 2002).

Similarly Bill Todd, a Tampa police detective, praised the closed circuit television (CCTV) system installed in Ybor City, Florida, specifically because, in his view, it is unable to discriminate on the basis of racial, ethnic, or gendered identity. Ybor City's biometric system uses facial recognition technology to identify suspects by using smart CCTV cameras installed in the city's Latin Quarter. Todd asserted that one of the primary benefits of installing the thirty-five cameras was that, unlike the police force itself, smart CCTV cameras are "race and gender neutral" (Lewine 2001). The allegedly arbitrary choice of this particular test site for facial recognition technology is problematized by Kelly Gates (2004), who highlights the tensions associated with choosing this neighborhood to test out a new biometric profiling technology given its raced and classed identity.

Claims of biometric neutrality are codified in book-length technical briefs (Pugliese 2005; Murray 2009). In their guide to the development and use of biometrics for businesses, John Woodward, Nicholas Orlans, and Peter Higgins (2003) argue that biometric technologies are beneficial because of their lack of bias: "The technological impartiality of facial recognition offers a significant benefit for society. While humans are adept at recognizing facial features, we also have prejudices and preconceptions." Woodward and his collaborators offer contemporary controversies concerning "racial profiling [as] a leading example." Contrasting human recognition with biometric recognition, they argue that facial recognition technology is incapable of profiling based on identity because "facial recognition systems do not focus on a person's skin color, hairstyle, or manner of dress, and they do not rely on racial stereotypes. On the contrary, a typical system uses objectively measurable facial features, such as the distances and angles between geometric points on the face, to recognize a specific individual. With biometrics, human recognition can become relatively more 'human-free' therefore free from many human flaws" (254).

Biometrics companies sell their technologies on the strength of these claims. For example, AcSys, a Canadian company dedicated to developing biometric facial recognition technology, promotes their product as "completely race independent—eliminating risk of racial profiling" (AcSys Biometrics Corp. 2007). Even if companies acknowledge that one type of biometrics, such as finger imaging, is not "race neutral," they suggest that a different biometric technology is the solution to this problem. For example, Doug Carlisle is a board member at A4Vision, a biometric company dedicated to developing 3D facial-imaging technology. He asserts that the bias associated with biometric fingerprinting can be fixed by using facial recognition technology: "Fingerprint technology . . . is difficult to implement due to physiological differences across varying ethnic groups as well as some cultural prejudices that resist fingerprinting. Using the shape of one's face and head is less invasive, more accurate and the most promising going forward for identification purposes" (Sheahan 2004).

Assumptions concerning the ability of biometrics to work with mechanical objectivity or within frameworks of "knowledge engineering" (Lynch and Rodgers n.d.), in which scanners eliminate subjective human judgment and discrimination, have made their way from media and scientific reporting into commonsense assertions about the neutrality of biometric technologies. Thus in an online discussion on the use of iris scanners at the U.S.-Canada border, one discussant claimed he would prefer "race-neutral" biometric technologies to racist customs border officials: "If I was a member of one of the oft-profiled minorities, I'd sign up for sure. Upside—you can walk right past the bonehead looking for the bomb under your shirt just because of your tan and beard. . . . In short, I'd rather leave it up to a device that can distinguish my iris from a terrorist's, than some bigoted lout who can't distinguish my skin, clothing or accent from same" ("Airport Starts Using Iris Screener" 2005). Here we see what Sherene Razack (2008) terms "race thinking," that is, that the state can justify the surveillance and suspension of the rights of othered communities in the interests of national security. The belief is that new technologies will circumvent forms of "race thinking" and racial profiling by replacing the subjective human gaze with the objective gaze of the state.

Institutions investing in biometric technologies use these assumptions about the technical neutrality of these new identification technologies to justify their adoption. These claims are interesting in a number of ways, not least in that they reveal tacit assumptions around race and gender discrimination by those who might not ordinarily acknowledge prejudice in the systems of which they are a part. It is relatively rare to hear a police officer argue for new technologies because his own task force is prone to racist and sexist assumptions, as did Bill Todd of Ybor City (Lewine 2001). Institutional players adopting biometrics represent them as the answer to institutional forms of discrimination and inequality such as racial profiling. And yet biometrics have failed, calling into question industry and scientific assertions about the objectivity of these new identification technologies.

Biometrics Unraveling

At the unveiling of a new iris scanner at Edmonton International Airport, Deputy Prime Minister Anne McLellan moonlighted as public relations manager for the biometric industry. McLellan demonstrated the operation of a new scanner designed to showcase the accuracy and efficiency of these new identification technologies and to promote CANPASS,[2] a program that replaces border personnel with biometric scanners. As we will see in greater detail in chapter 4, CANPASS is a NEXUS-Air affiliate border-crossing program designed to speed "pre-approved, low-risk air travelers" across the border. Meant to showcase the efficiency of replacing subjective humans with objective machines, the demonstration did not go as planned: "McLellan stared into the scanning machine . . . but twice a computerized voice declared: 'Iris scan unsuccessful'" ("Canada's Edmonton Airport Implements Iris-Recognition Screener at Immigration" 2011).

As the Edmonton airport failure reveals, new technologies designed to ease identification do not always function in the straightforward way that industry discourse would have us believe. News of biometric failures circulated shortly after their introduction, as test results indicating their poor efficacy quickly made their way into the media. Simultaneously indications of the ease with which they could be "spoofed" (fooled into malfunctioning) proliferated. Before turning to the science on which biometric technologies rely, it is useful to examine the cracks in the façade of biometric success. That is, it is helpful to look at reported biometric failures before investigating why these errors might occur.

Surveys, Spoofs, and Hacks

One of the earliest documented failures of a biometric identification system involves facial recognition technology used in Ybor City, Florida. Given the significant costs and possibilities for surveillance provided by the system, and given its use within a predominantly Latino neighborhood, the system attracted significant media attention from its inception. In 2003 the American Civil Liberties Union documented that, after its first year, the system worked so poorly that the police were unable to identify a single criminal using a biometric facial recognition system that included thirty-five cameras. Although the system was renewed for a second year, it proved so difficult, expensive, and time-consuming for the police to use, while yielding only mismatches and failing to recognize known subjects, that it was ultimately abandoned (Gates 2004).

Other reports of biometric failures soon followed. The ACLU (2003) obtained documentation concerning a pilot project testing facial recognition technology at Logan Airport in Boston in which the photographs of forty employee volunteers were scanned into a database. The employees then tried to pass through two security checkpoints equipped with biometric cameras. The volunteers could not be identified 96 times out of 249 over a period of three months, a 39 percent failure rate. Key to these failures were lighting distinctions between the original photograph and the image captured on the biometric cameras. A security expert, Bruce Schneier (2001), suggested that technological dependence on good lighting conditions might be problematic, given that it is unlikely that terrorists will stop to pose for well-lit photographs. More recently the Face Recognition Vendor Test (FRVT), conducted in 2006 and partially sponsored by the Department of Homeland Security and the FBI, concluded that the error rate for facial recognition technology has decreased since 2002, partly because the technology is better able to identify faces across different lighting environments. The ideal was set as a false rejection rate and a false acceptance rate of 1/1000. In 2002 one out of five people was not accurately identified; by 2006 one out of a hundred was not accurately identified. However, an error rate of 1/100 is still very high. Given that the International Civil Aviation Organization standard requires that all passports eventually contain facial biometric information, and given the volume of traffic that passes through airport security, this represents a significant number of people who will encounter difficulty (P. J. Phillips et al. 2007:5). At Logan International Airport, which handles 27 million passengers a year, 739 people would not be correctly biometrically identified on an average day.

Even those biometric technologies that did not fail were only as strong as the privacy measures in place to protect individual biometric information. There have already been a significant number of security breaches. At a convention intended to showcase new biometric technologies, the fingerprint and retinal information of thirty-six people trying out vendor products accidentally was sent by email to everyone attending the conference (Williams 2008).

The number of reports about the ease with which high-tech biometric technologies are hacked also calls their security into question. Researchers at Yokohama National University found that biometric fingerprint readers are easy to deceive using artificial gelatin fingers onto which a real print was dusted (Matsumoto, Matsumoto, Yamada, and Hoshino 2002). In Germany the magazine *c'T* published extensive results on ways to successfully spoof biometric technologies (Thalheim, Krissler, and Ziegler 2002). Their methods included using high-quality digital printouts of irises to deceive iris scanners, using a little water on a biometric fingerprint scanner in order to reactivate the previous person's latent image, and moving one's head slightly from side to side to outwit facial recognition technology scanners. Marie Sandström (2004), a scientist, reproduced *c'T*'s success in deceiving biometric fingerprinting; she tested nine different biometric fingerprint recognition systems and found that each was fooled by an artificial gelatin imprint.

As I noted in the introduction, one kind of biometric fraud causes particular anxiety. The possibility that a biometric scanner can be deceived by a severed body part garnered a good deal of attention, as was apparent by the multiple references to this possibility in mainstream news reports and media representations (Geoghegan 2005; Richtel 1999; Carney 2007; Mullins 2007). Moreover ongoing scientific research on the best method to prevent a biometric scanner from recognizing the body part of a corpse (Abhyankar and Schuckers 2006; Parthasaradhi et al. 2005; Antonelli et al. 2006) gives credence to contemporary anxieties about the ramifications of biometrics for bodily security.[3]

High-Tech Racism

Biometric technological failures do not occur equally across the board. The growth of biometrics occurred in a climate of increased anxieties about the "inscrutability" of racialized bodies. Given this cultural context, it is unsurprising that reports that biometric technologies did not function in the objective ways claimed for them began to proliferate. For example, a business brief about the possibilities of using biometrics for secure transactions reported that some "Asian women . . . had skin so fine it couldn't reliably be used to record or verify a fingerprint" (Sturgeon 2005). Advocates of multimodal biometrics, in which more than one technology is used, took as their starting point that biometrics had difficulty recording the image of the bodies of people of color. Multimodal biometrics claimed to be able to enroll more othered bodies than could a unimodal technology: "At first glance, a Multi Modal biometric system offers certain advantages for your business, when contrasted to a Uni Modal security system: A Multi Modal system can capture the unique, biometric characteristics of a much larger and more varied target population. For example, it has been known that certain ethnic races have more difficulty in enrolling and verifying than other ethnic races. A Multi Modal system can help alleviate this problem over time" (Das 2007). Similarly the U.S. Transportation Security Administration (TSA) claimed they used a biometric iris system instead of one dependent on biometric fingerprinting because an iris scanner "allow[s] broader participation," whereas "gender, ethnicity and age affect the qualities of fingerprints" (Fletcher 2006:20).

Despite TSA's claim about the objectivity of iris scanning, reports circulated about iris scanners that functioned differentially on racialized bodies. Media reporting on biometric trials in the U.K. claimed that iris scanners disproportionately were unable to read the irises of people of color (Saeb 2005). Other news reports suggested that racism was being technologized in the form of the biometric iris scanner. It seemed that the darker a person's skin, the greater the technological failure: "Answering a question on whether it regarded the rates as satisfactory, the Home Office replied: 'It is true to say that at times in the past some difficulties have been experienced in successfully recording the iris images of people with very dark skin (on some iris systems). The difficulty lay in the ability of the system to successfully locate irises against a darker skin tone' " (Kablenet 2006). This response fails to explain why skin color would make a difference, as the irises of all people are surrounded by the white of the eye.

Such examples point to the problematic failures of biometric technologies. Activists fighting the implementation of biometric ID cards raised alarms about biometric plans, saying, "Little did we know that the government's own computers would actually be favoring blue eyes" (Saeb 2005). The ways that biometric technologies best imaged bodies racialized as white is reminiscent of Richard Dyer's (1997) work on how film was developed in order to be able to visualize whiteness. Evidence suggests that both government and industry representatives promoting biometrics were well aware of the differential success of biometric identification technologies in the accurate recognition of racialized persons, as news reports of biometric trials in the U.K. revealed that the technologies were tested on different racial and ethnic communities. These tests suggested that the government expected that different groups would yield different results: "Volunteers [will] be recruited from different ethnic groups to help . . . analyze the results [and] to examine any differences in results by demographics" (Kablenet 2006). Testing biometric technologies on different communities in this fashion poses an interesting contradiction. It undermines earlier industry assertions that biometric technologies are race-neutral while simultaneously recognizing the material effects of racial categories.

The technological fallibility of biometrics did not end with racialized bodies. News reports that biometric technologies consistently could not identify those deviating from the "norm" of the young, able-bodied male suggested that, in general, "one size fits all" biometric technologies failed to work. Here we see the perils of what Haraway (1997:142) terms "corporeal fetishism," as biometric science is imagined to be able to condense complex relationships and situated knowledges into a single digital map of the body, one free from the "failures" of culture and understood to be outside troping or representational strategies. In the process of corporeal fetishism, particular bodies are rendered legible to new technologies and then packaged into a commodified form able to be circulated in the transnational networks of global capital as part of what Simone Browne (2009) calls the "identity-industrial complex." Here we have the failure of biometric science to recognize bodily complexity. These types of biometric failures result in part from what Haraway (1997:137) would call the error of mistaking complex and "worldly bodies-in-the-making" for fixed and reified things. A news article posted on Black Information Link, a community website by and for people of color living in the U.K., suggested that biometric facial recognition technology worked very poorly with elderly persons and failed more than half the time in identifying those in the study who have disabilities (Saeb 2005). Other news articles suggested that biometric iris scanners were particularly bad at identifying those with visual impairments and those who are wheelchair users, due to difficulties in getting the angle between the eye and the iris scanner within the parameters required for the scanner to work or because people could not see the red light in the scanner itself directing them where to look (Gomm 2005). Class also was a factor, as hard physical labor erases fingerprints. The U.K. Biometrics Working Group (2002) found that accurately fingerprinting those with "clerical, manual, [and] maintenance" occupations was more difficult than fingerprinting those in other occupations. Biometric iris scanners additionally failed disproportionately with very tall persons (Gomm 2005), and biometric fingerprint scanners couldn't identify 20 percent of those with nonnormative fingers: "One out of five people failed the fingerprint test because the scanner was 'too small to scan a sufficient area of fingerprint from participants with large fingers' " (Saeb 2005). In short, any kind of bodily deviance could give rise to biometric failure, begging the question as to exactly what type of body would not yield errors. "Worn down or sticky fingertips for fingerprints, medicine intake in iris identification (atropine), hoarseness in voice recognition, [and] a broken arm for signature" all gave rise to temporary biometric failures; "well-known permanent failures are, for example, cataracts, which makes retina identification impossible or rare skin diseases, which permanently destroy a fingerprint" (Bioidentification 2007). In a blatant acknowledgment that their biometric scanners failed to work for people with a number of disabilities, the Clear program in the U.S., which uses biometric iris and fingerprint information to speed travelers through security, posted the following warning:

> Our verification kiosk currently includes a scanning device that requires that you walk into and stand on the kiosk. Therefore, the kiosk may not be able to serve all our members.

> If you have any of the following conditions, please inform the Clear [program] attendants BEFORE entering the kiosk: artificial limbs, bone or joint replacements, metal implants, plates, rods, pins or screws, shrapnel, surgical staples or wires, pacemaker or other life-sustaining medical device(s).
>
> Clear is committed to working with TSA and our technology partners to develop and deploy a registered traveler solution that serves the broadest possible population. (Clear Program Guidelines 2007)

Despite its token attempt to suggest that it works with a broad population, Clear is one among many examples of biometric programs that are unable to serve customers with disabilities. The Clear program helps to reveal which bodies can be rendered as commodities and which cannot, suggesting that disabled bodies cannot be rendered as financially profitable objects. This is also true of the biometric iris scanners used to speed travelers across the U.S.-Canada border: you must be able to stand and you must have relatively normative vision to use the NEXUS-Air biometric scanners; thus wheelchair users and those with visual disabilities may not participate in this program. Here we are reminded of the ways that corporeal fetishism is used to transform bodies into objects—ascribing a value to every body—as some are deemed to have more economic value than others.

Assertions regarding the inability of biometrics to identify othered bodies also could be found in biometric textbooks (Murray 2007; Pugliese 2005). In their primer, *Biometrics: Identity Verification in a Networked World*, Nanavati, Thieme, and Nanavati (2002:60) explain one of the weaknesses of finger scans: "Certain ethnic and demographic groups have lower-quality fingerprints and are more difficult to enroll than others. IBG's Comparative Biometric Testing has shown that elderly populations, manual laborers, and some Asian populations are more likely to be unable to enroll in some finger-scan systems." They describe similar difficulties with iris-scanning technology for people with disabilities: "Users with poor eyesight and those incapable of lining up their eyes with the technology's guidance components have difficulty using the technology" (85). Given the long list of reasons why a biometric scanner might fail, it seems that these technologies work best for blue-eyed males with good eyesight and no disabilities, neither too young nor too old, a Goldilocks subject who is "jussstright." The group that can be reliably identified is a relatively small percentage of the population, though perhaps a somewhat more significant proportion of business travelers and other users of voluntary biometric programs.

Although technological utopianism and biometric technologies as a panacea for social problems are a key part of the fabric of the discourse on these new surveillance technologies, we see that, rather than reciting a narrative of seamless technological functionality, the biometric failures I have detailed, including their disproportionate failure on othered bodies, suggest that these technologies do not operate with the mechanical objectivity claimed for them.

Scientific Representations of Biometrics

Everyone always wants the latest and coolest stuff, and this holds true for the computer industry. Many firms eagerly jump into the next big technology so they can say they were there first. This is referred to as being on the leading edge. In recent times, many have believed that biometrics were not on the leading edge, but on the bleeding edge. The bleeding edge is where a technology gets debunked. Biometrics have spent more time than any other technology in recent memory on the bleeding edge.

PAUL REID, *BIOMETRICS FOR NETWORK SECURITY*

Most vendor statements on accuracy bear little relevance to real-world performance of biometric systems.

SAMIR NANAVATI, MICHAEL THIEME, AND RAJ NANAVATI, *BIOMETRICS: IDENTITY VERIFICATION IN A NETWORKED WORLD*

The number of reported failures in biometric technologies demonstrates the fallibility of any technology that takes as its starting assumption the consistency of bodily identity. As we saw, the biometric industry regularly asserts that any difficulty with these technologies can be solved with sufficient research, money, and time. Although billions of dollars have been injected into state-funded biometric programs—from the prison system to the welfare system to border security—persistent failures remain. Like the polygraph, a technology whose perfection we continue to be assured is just around the corner (Littlefield 2005), it remains far from clear that biometrics will ever work reliably, no matter how long we leave them to gestate in the publicly subsidized private labs of biometric companies. It is necessary to investigate those instances when biometric technologies fail and to ask what their failures tell us. The moments in which they fail are useful to identify the assumptions upon which they rely and the cultural context that they encode.

I turn now to an investigation of scientific studies on biometric technologies designed to improve their ability to classify individuals based on bodily identity.[4] Biometric scientists rely upon the computerized recognition of gender and race to help with the development of significantly faster biometric identification—one of the vaunted goals of the industry. In general these studies argue that it is useful to reduce the size of the database against which a particular individual is matched, especially when performing a one-to-many (1:N) search, and they rely upon gender, racial, and other categories of identity to reduce the population size (the N) against which the individual is screened. If the person being identified is already classified as male and thus need only be checked against a database of other males, the argument is that this will cut computer processing time in half (Childers et al. 1988:603). As well, using so-called soft biometric traits like gender, race, and age in order to classify subjects is hypothesized to enhance the overall accuracy of biometric recognition (Buderi 2005; Jain, Dass, and Nandakumar 2004b). Soft biometrics "are nondistinctive, limitedly permanent human traits such as gender, age, hair color, weight and height" (Jain, Dass, and Nandakumar 2004a).

Much of the research on biometric recognition is based on a foundational article titled "SexNet: A Neural Network Identifies Sex from Human Faces" (Golomb, Lawrence, and Sejnowski 1991). A research team led by the biologist and medical doctor Beatrice Golomb combined developments in computing and informatics with rudimentary biology in order to develop a computerized network able to identify sex from frontal facial photographs. Giving their network the somewhat obvious name "SexNet," the team concluded that their machine could classify human faces by sex more reliably than people could. They postulated that the computer network's average error rate was 8.1 percent, in contrast to 11.6 percent for human error. However, SexNet's overall error rates remained high. One in ten faces was incorrectly identified. The researchers found some faces especially problematic for both humans and computers. One male face was mis-sexed by both, suggesting perhaps that human difficulties in identifying biological sex from faces were being technologized. As stated by Golomb, Lawrence, and Sejnowski, "Similar errors seemed to affect the net and humans. One male face gave particular trouble to SexNet, being mis-sexed when a test face, and taking long to train when a training face. This same face was (erroneously) judged 'female,' 'sure' by all 5 human observers" (575). In order to address this difficulty, the researchers suggested creating a "special category for the individual" and others like him (575). In this way one would avoid needlessly amending the male and female categories in order to be able to correctly classify what was referred to as one "fluke" face (575). This would additionally permit sex "outliers to be correctly identified without adverse consequences to generalization" (576). Thus individuals whose biological sex was not immediately apparent to either people or computers, a sample imagined by Golomb, Lawrence, and Sejnowski to be a tiny fraction of the human population, would not interrupt SexNet's ability to classify the vast majority of faces.

In the event that readers of the study should think that these sex "outliers" indicated that SexNet functioned poorly, Golomb, Lawrence, and Sejnowski cited the following example of the network's superior ability to identify sex as proof that their system worked. During the labeling of the sex of each photograph, one female face was accidentally assigned a male value. This human error was detected by SexNet, as described in the results section of their paper: "SexNet proved right: The face was a clear female whose sex value had been mistranscribed" (1991:575). This example was given as evidence that the network "had evidently done a fine job of abstracting what distinguishes the sexes" (575) and was able to eliminate problematic human intervention in the pure practice of classification. A number of future applications for SexNet were outlined in the study's conclusion, including the suggestion that this network could help to indicate whether sex could be extrapolated from the face of other mammals in addition to humans, for example, an experiment investigating whether the faces of rhesus monkeys differed by sex. Golomb and her collaborators suggested that a second, more "frivolous" application for systems like SexNet would be to "scientifically test the tenets of anthroposcopy (physiognomy), according to which personality traits can be divined from features of the face and head" (576). In this way the development of biometric systems of classification attempted to reopen discussions about the scientific worth of formerly discredited branches of scientific inquiry such as physiognomy. This type of biometric research also marks the return to contemporary biological understandings of gender and race, a debate that is once again on the rise.

Though these technologies may fail to work in the ways claimed for them, the case studies in this book show how biometric technologies work to exclude particular communities, including prison inmates, welfare recipients, and immigrants and refugees. Applications for this type of biometric research are abundantly clear. No longer restricted to an underfunded future sorting rhesus monkeys, biometric technologies were rolled out as an essential part of the security-industrial complex. Moving from the biological to the computational sciences, studies on ways to refine SexNet proliferated. In their study of how to classify facial images by gender, Jain and Huang (2004) suggested using two algorithms to extract geometric features and classify human faces. In "An Experimental Study on Automatic Face Gender Classification," Yang, Li, and Haizhou (2006) attempted to use a "normalization" process to classify frontal facial images by gender. Running photos of faces through a number of statistical procedures before generating their digital template, they compared three algorithms in order to see which better classified their database of normalized faces. The database consisted of "11,500 Chinese snapshot images, including 7000 male and 4500 female . . . whose faces are always upright and neutral without beard[s] or strange hair style[s]" (3).

In this study, Yang, Li, and Haizhou concluded that subjecting the photos to statistical normalization helped the computer to learn to classify the faces correctly. It appears from figures 2.1 and 2.2 that smiling is a telling indicator of femaleness, in keeping with other studies that conclude that showing emotion in the face is a key indicator of female identity. Also, the face with long hair is identified as female, whereas the face with short hair is identified as male. One of the normalization procedures Yang, Li, and Haizhou followed was to run all of the photos through a statistical method of triangulation warping, in which the shape of the individual face is discarded in favor of a mean face shape. What Golomb, Lawrence, and Sejnowski (1991) called "sex outliers" were shed in the move from biology to computer science. In keeping with homophobic ideologies that assert that God made Adam and Eve, not Adam and Steve, these later studies (Jain and Huang 2004; Ueki et al. 2004; Givens et al. 2004; Yang, Li, and Haizhou 2006) rejected Golomb's "sex outlier" category altogether. Rather than outliers, these studies concluded that faces that could not be recognized by the machine were evidence that the technology needed further fine-tuning.

Some sex classification systems attempted to use biometric indicators other than the face in order to categorize their subjects. One early study

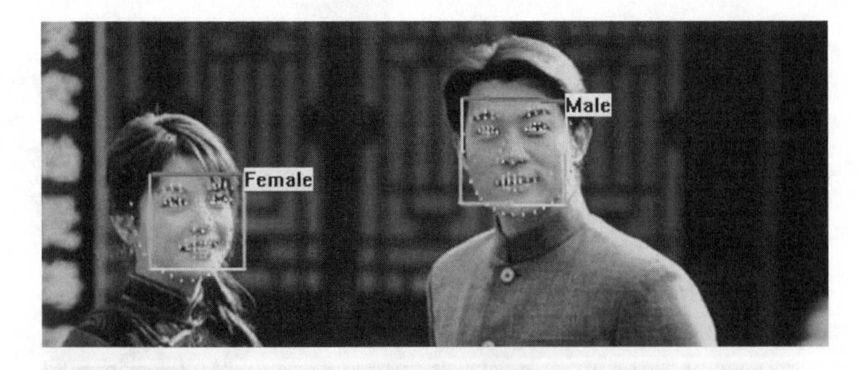

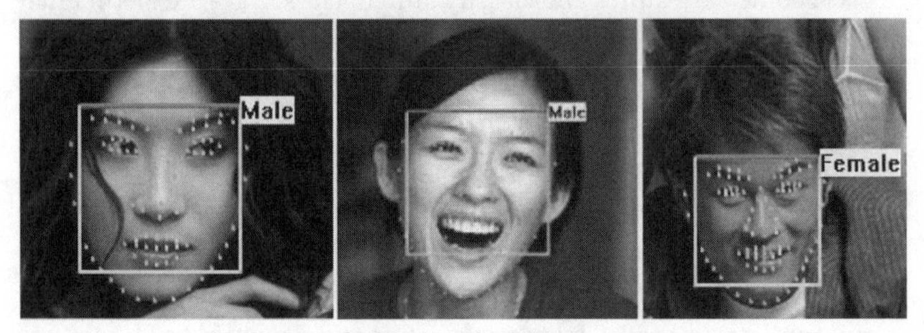

FIGURES 2.1 and 2.2 In the top row are faces whose gender was correctly identified; in the bottom row are faces whose gender was not correctly identified (Yang, Li, and Haizhou 2006). Courtesy of Haizhou Ai.

used voice biometrics to classify individuals by sex. Childers et al. (1988) claimed to be able to identify the sex of their subject based on the number of voiced fricatives, a sound produced by pushing air through a restriction in the vocal tract, such as occurs when producing the sounds "f" and "s." They concluded that women uttered fricatives more commonly and in this way could be differentiated from men (603). Other studies used skin color biometrics in order to improve recognition. In these cases the color of the skin of one individual was superimposed on a different photograph of the same individual, with the idea that this would make the individual more identifiable over time (Yin, Jia, and Morrissey 2004; Li, Goshtasby, and Garcia 2000). In their study on biometric classification by gender, Ueki et al. (2004) attempted to demonstrate the importance of sartorial markers to consistent bodily identity. They tried to show the importance of "neckties and décolletés (clothes with [a] low-cut neckline)" to "differentiate gender" (see figs. 3 and 4).

In addition to clothing cues, Ueki et al. attempted to integrate the subject's hairstyle into the process of biometric classification, asserting

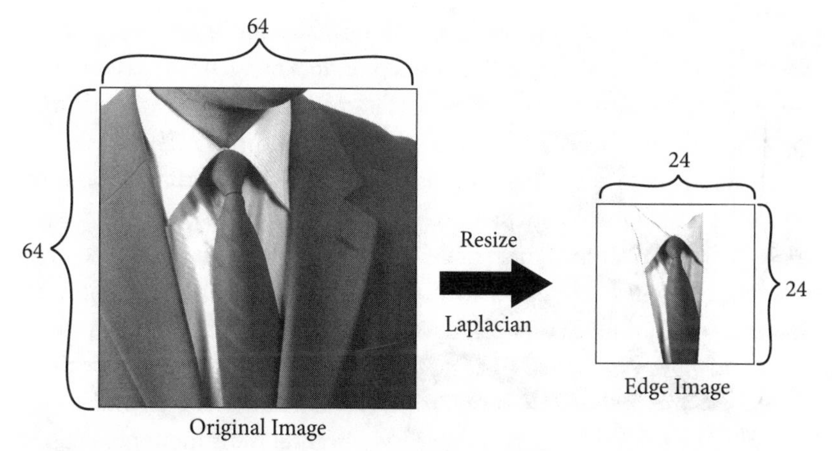

FIGURE 3 The biometric template of the necktie is created from the raw biometric data (Ueki et al. 2004). The Laplacian is a scalar operator that can transform an image into an image more easily analyzed.

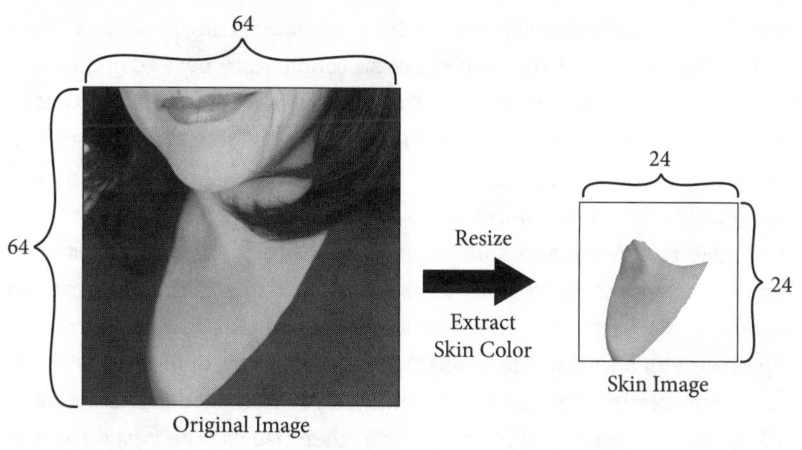

FIGURE 4 The biometric template of the neckline is created from the raw biometric data (Ueki et al. 2004). The skin image is meant to demonstrate whether the neckline is revealing (female) or discreet (male).

that "hairstyle is considered to be one of the effective features in gender classification" (fig. 5). Postulating that long hair helps to demonstrate that the unknown subject is a woman, whereas short hair indicates masculinity, Ueki et al. concluded that neckties, hairstyles, and whether or not the individual is wearing a revealing top all improve the system's ability to classify individuals by gender. As a quick glance could accomplish the same task for a lot less money than would be required to refine and

FIGURE 5 How the biometric template of the hairstyle is created (Ueki et al., 2004).

implement this biometric system, we need to keep in mind the profit motives driving the development of these technologies.

The methods used for gender classification were next adapted to attempts to use biometrics to categorize subjects by race. One example is found in "A Real Time Race Classification System" (Ou et al. 2005). The researchers set out to design a cataloguing system that reliably could use biometric facial recognition technology to divide the scanned faces into two categories, Asian and non-Asian. First, they proposed to turn the raw biometric data extracted from the face into a digital template using an algorithmic process they called principal component analysis (PCA). Using this method they located each face's eigenvectors, defined as "mathematical properties that describe the unique geometry of a particular face" (P. Reid 2004:99). Ou et al. (2005: 379) then used a supervised form of PCA to isolate the eigenvectors that contained key information about race. Using a technique called independent component analysis (ICA) they extracted the features most useful to classifying each face. They argued that this method would help to isolate the facial features with the best "race information" (380). Once they had refined their PCA and ICA processes to select and extract the features that contain the key information about the test subject's race, they then classified their facial images into two categories: Asian and non-Asian. Using a statistical technique called Support Vector Machines, they flattened each face to reduce the three-dimensional model to a two-dimensional model (380). Ou et al. then input a large database of faces into their computer and manually labeled the race of each face. Having "taught" the computer which faces represented which race, Ou et al. tested the computer on a new series of faces. They found their total error rate to be 17.5 percent.

The method followed in their paper is standard. Their major contribution is the PCA and ICA methods they used to generate the digital template of the face. The authors also proposed a new algorithm to improve their chances of selecting features that contain good race information and to reduce the chance that they will choose features that contain bad race information (Ou et al. 2005:381). Other than the digital template contributions, the authors follow a well-defined path to the binary classification of bodies into racial categories.

Other biometric studies attempted to analyze the impact that a particular race's "feature characteristics" might have on the ability of these individuals to be recognized by a given biometric technology. In a paper titled "Rapid Pose Estimation of Mongolian Faces Using Projective Geometry," Hua-Ming, Zhou, and Geng (2004) used anthropometry as the foundation of their study. Anthropometry arose out of physical anthropology and is the technique of reading human characteristics (including intelligence) off the body. It is a largely discredited branch of science, famously debunked by the Harvard biologist Stephen Jay Gould (1996). In 1996 the American Association of Physical Anthropologists issued a statement denying any biological theory of race. Yet, as we see in the paper by Hua-Ming, Zhou, and Geng, anthropometry as a field of study is being revived by biometric scientists, including Golomb, Lawrence, and Sejnowski (1991) in their suggestions of possible future applications for their sex classification system. The resurgence of interest in reified understandings of race and gender reflects the debates that continue to rage over whether race is biological, despite the wealth of scientific evidence to the contrary.

Hua-Ming, Zhou, and Geng (2004) begin their study of "Mongolian" faces by asserting, "The difference of Race is obvious, and it is the central field of the research of anthropology. Anthropometry is a key technique to find out this difference." They declare, "[As] the Mongolian race is regarded as the main race of China, the statistical information of its feature[s] has very important scientific meaning to our research in the field such as facial pose estimate and face recognition." These scientists predictably define the Mongolian race's distinctive features as the "width of the eye feature" and the "length of the nose," characteristics that they

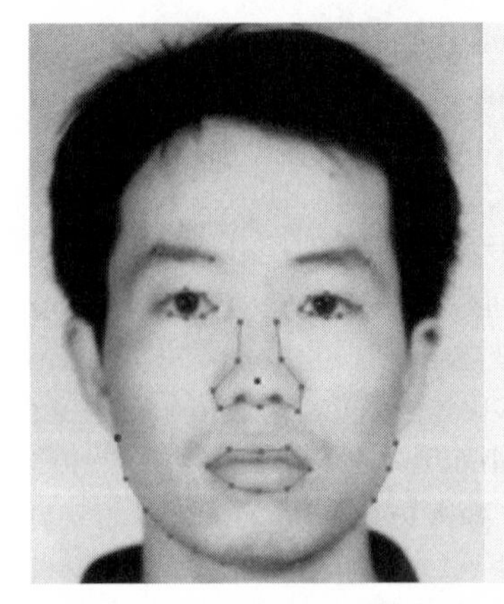
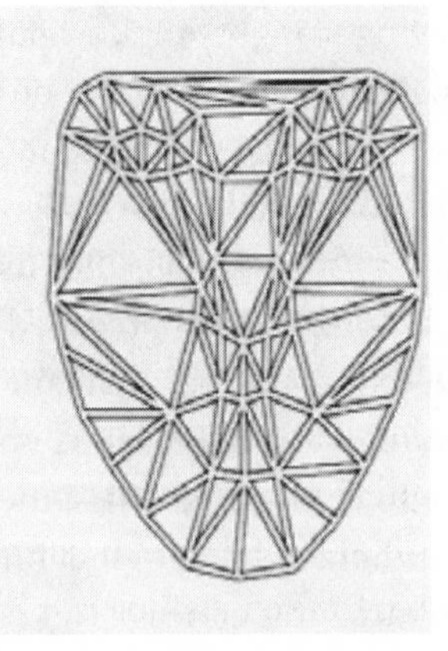

FIGURE 6 Biometrics and race: this image was taken from "Rapid Pose Estimation of Mongolian Faces Using Projective Geometry" by Hua-ming, Zhou, and Geng (2004). Eigenvectors are laid out in a grid over the mouth, nose, eyes, and chin. Courtesy of Li Hua-ming.

argue are uncontaminated by facial expression, which may make facial recognition more difficult (fig. 6). In developing this model, the authors concluded that they would be able to improve computer facial recognition by accounting for what they term "the Mongolian race's feature characteristic." That is, in refining biometric technologies to identify Mongolian faces, Hua-ming, Zhou, and Geng would help facial recognition technology work well in China.

Many of the studies attempting the biometric classification of bodies aim to categorize individuals by both gender and race. For example, in their study Givens et al. (2004) conducted a comparison test of the efficacy of three algorithms at identifying faces based on their features. They included gender and race as facial features, alongside "glasses use, facial hair, bangs, mouth state, skin complexion, state of eyes, makeup use, and facial expression." In indicating how to classify each individual, Givens et al. give the following explanation of their categories:

> Race [White*, African American, Asian, Other.] The "Other" category was used for Arab, Indian, Hispanic, mixed race, and any subject that did not fit into the other three categories.
>
> Gender [Male*, Female.] Self-explanatory

The asterisk indicates the factor that was designated as the baseline; that is, it was assumed as the default category. While the scientists conducting this study do acknowledge that it is possible for the rating of each facial feature to have a subjective component, they stress that it is not statistically significant. They conclude, "Gender, skin, glasses and bangs were easy for our viewer to judge, [while] race, facial hair, expression, mouth and eyes were somewhat harder, although the viewer was still confident in his judgments." When the judge could not decide, the default values of white and male were selected automatically.

Givens et al. ultimately found that, regardless of the algorithm used, "old [elderly] subjects and non-Caucasian subjects are easier to recognize." For one of the algorithms, women are harder to identify, whereas for the other two there is no significant difference. Although the authors asserted that most of the work on gender classification finds that men are easier to recognize they cite no sources. Givens et al. attributed the improvement of their study on other attempts at biometrically recognizing women to the fact that they also controlled for factors such as facial hair and makeup.

Givens et al.'s classification of faces based on features is just one study among many that attempt to group subjects according to both their gender *and* race identities. Invariably these categorization processes are done sequentially rather than simultaneously. For example, in their study "Gender and Ethnic Classification of Face Images" Gutta, Wechsler, and Phillips (1998:194) developed two classification tasks, one for gender and one for race. This type of sequential ordering is repeated in "Mixture of Experts for Classification of Gender, Ethnic Origin and Pose of Human Faces" (Gutta et al. 2000) and "Classifying Facial Attributes Using a 2-D Gabor Wavelet Representation and Discriminant Analysis" (Lyons et al. 2000). In all of these studies "hybrid" subjects (for example, women of color) would need to be classified first by gender and then by race, though no instructions for how to classify these subjects are provided. This is especially ironic given that the new racial categories used in the U.S. census of 2000 showed increasing numbers of "hybrid" individuals selecting two or more racial categories. One might ask about the difficulties that individuals like President Barack Obama would face under this biometric regime.

These experiments represent only a few of the many biometric studies that classify individuals based on bodily identity, each using slightly different methods or a different data set. The initial labeling of the images by gender or race in order to train the computer system must be done manually by the scientists themselves. The following statement is standard, demonstrating that so-called physiognomic differences are key: "The ground truth label for gender and ethnic origin were determined by visual inspection after the images were collected" (Gutta et al. 2000:198). That is, scientists themselves decide on the gender and race of an individual before using algorithms to train their computer to do the same. While the biometric classification method in these studies remained constant, the categories themselves varied. Individuals may be labeled Asian or non-Asian (Ou et al. 2005), Mongolian or non-Mongolian (Hua-Ming, Zhou, and Geng 2004), East Asian or non–East Asian (Lyons et al. 2000), Japanese or non-Japanese (Tanaka et al. 2004), Caucasian, Asian, Oriental, African, South Asian, or East Asian (Gutta, Wechsler, and Phillips 1998; Gutta et al. 2000). The ways that biometric technologies identify racialized and gendered bodies differently is known to scientists. From comparisons of the impact of biometric scanners that could identify some "races" better than others (Tanaka et al. 2004) to attempts to teach biometric systems to classify gender (Ueki et al. 2004), these studies help to explain biometric failures and their connection to race and gender identities. The papers just reviewed rely on racist research that is long debunked, while ignoring the host of empirical work in the past century on the complexity of bodily identity.

Significance: Historical Work on the Human Body

The scientific papers I have discussed rely predominantly on the fields of computer processing and recognition and largely ignore the rich field of empirical research that arose from studying human expression, emotion, and movement. Studies on the complex nature of the human face date back to 1806, with the publication of "Essay on the Anatomy of Expression in Painting" (C. Bell). More recent research includes the work of Ray Birdwhistell, one of the founding fathers of kinesiology, who spent much of his career attempting to develop a classification system for nonverbal communication. Other work in this area includes the research of Paul Ekman, whose book *Emotion in the Human Face* (Ekman, Friesen, and Ellsworth 1972) provides a detailed examination of the relationships between emotion, gender, age, and facial expression, and upon whom the TV show *Lie to Me* claims to be based. In the introduction to his book Ekman demonstrates the complex nature of human facial expression with his assertion that "man's facial muscles are sufficiently complex to allow more than a thousand different facial appearances; and the action of these muscles is so rapid, that these could all be shown in less than a few hours' time" (1). The rich literature on acting also gives detailed descriptions of the complexity of the human body and how to portray a great variety of human expressions, calling into question any notion of an essential or immutable bodily identity. Given this literature on the complexity of the human body, it seems strange that biometric scientists rely upon long-discredited assumptions about the biological nature of identity while refusing to engage with a field of research that dates back more than a hundred years. Certainly the omission of even a cursory review of this vast field of research seems egregious. Biometric discourse clearly functions as a form of corporeal fetishism in which complex human bodies are represented by ones and zeroes without discussion of the bodily processes that these strands of binary code hide from view or the implications of this particular representational strategy.

Biological Understandings of Race and Gender

The studies I analyzed earlier reveal some of the reasons that biometric technologies regularly fail, and why race and gender identities are so often connected to these failures. We saw that many biometric technologies are founded upon erroneous understandings of race and gender as reified phenotypes. In this way biometrics remind us of earlier technologies that attempted to read deviance off the body, including craniology, phrenology, and anthropometry. Biological categories of race have long been debunked by scientists (Tishkoff and Kidd 2004; Wells 2002; Kaufman 2008), social scientists (Sternberg 2007; Fish 2000; Duster 2005; American Association of Physical Anthropologists 1996), and cultural theorists (Gould 1996; Hammonds 2006). Similarly the biological nature of gender is problematized by a wide range of fields, from biology to philosophy to women's studies (Kessler 1990; Haraway 1997; Fausto-Sterling 2000; Butler 1990). Many of the biometric papers I cited refer to much earlier work that attempted and failed to definitively measure race and gender.

Consider the following quotation from an atlas of electroencephalography (the neurophysiologic measurement of the electrical activity of the brain, or the measurement of brain waves), written in 1941: "It is possible to tell an Eskimo from an Indian by the mathematical relationship between certain body measurements at a glance. . . . It would be wrong, however, to disparage the use of indices and objective measurements;

they are useful and should be employed wherever possible" (Gibbs and Gibbs 1941:n.p.). The study of racial classification and the use of biometrics by Ou et al. (2004:379) is more than slightly evocative of that earlier study: "We consider that every [facial] feature encodes different information in a certain scale such as gender, race, identity, etc. . . . By analyzing the distribution of the project 2N points, we can roughly select the eigenvectors which have more race information." Similarly, the study by Gutta, Wechsler, and Phillips (1998:195) on the use of facial geometrics to classify subjects by race reminds us of the 1941 study. Although Gutta, Wechsler, and Phillips use computers, the narrative bears more than passing similarity to the earlier study:

> Face processing first detects a pattern as a face and then will box the face. It proceeds to normalize the face image to account for geometrical and illumination changes using information about the box surrounding the face and/or eyes location, and finally it identifies the face using appropriate image representation and classification algorithms. The result reported later on assumes that the patterns corresponding to face images have been detected and normalized. The specific task considered herein is that of gender and ethnic classification, i.e., discrimination of human faces as belonging to either a female or male category and Caucasian, Asian, Oriental or to African categories respectively.

In contending that particular races have associated characteristics or in suggesting that some body parts contain more race information than others, long debunked biological theories of race again take on the authority of objective scientific assertions. This ignorance of basic contemporary concepts—for example, that race is not biological even if systemic racism means that race as a category has material significance—gives rise to a host of difficulties. One ramification is that mixed-race individuals, in addition to the overwhelming number of people who do not meet the rigid racial types defined by these experiments, are ignored altogether by biometric classification systems as they do not easily fit defined categories, including those mixed-race individuals who get classified within groups. As a result of biometric science, entire communities become what Leigh Star and Geoff Bowker have called "boundary objects," Donna Haraway has called "monsters," and Kal Alston has called "unicorns"—uncategorizable objects that do not fit into any classification system, objects at once "mythical and unknowable, straddling multiple worlds"

(Bowker and Star 1999:306). Although systemic forms of discrimination give race and gender material significance, this does not imply that race and gender are stable qualities that can be reliably read off the body.

Nor do mixed-race individuals present the only problem with premising biometrics on biological understandings of race, as racist ideologies are necessarily at the core of any assertion that race can be reliably measured or classified. This helps to explain why biometric studies consistently feature whiteness and maleness as normative. For example, in using white and male as the default categories for biometric selection (Givens et al. 2004), or in attempting to determine whether biometric scanners will be able to identify the "distinctive" faces of the "Mongolian race" (Hua-Ming, Zhou, and Geng 2004), researchers are suggesting that white faces may easily be recognized, whereas nonnormative racialized or female faces pose problems for biometric recognition. In his book *White* Richard Dyer (1997) shows that using technologies to privilege whiteness has a long history. Using biometric technologies to measure race, ethnicity, and gender, new technologies serve as disturbing reminders of older technologies underpinning racist policies. Moreover biometric scientists turn to essentialist understandings of race and gender at precisely the time that the debate on whether race is indeed biological again rises to prominence. In 2006 the Social Science Research Council launched a forum titled "Is Race Real?" in order to problematize the causal connections being drawn between race and genomics. Assertions about the biological nature of race are not the province of low-profile science or scientists. The Nobel prize–winning molecular biologist James Watson asserted in 2007 that injections of melanin would increase sex drive, and claimed that black people did not have the same capacity to reason as white people (Milmo 2007).

Biometric systems of categorization have the potential to exclude both those who don't belong to the privileged category and those whose body is illegible. If you cannot obtain a passport unless your fingerprints can be collected—a proposal currently under consideration in Canada and the United States—then those without legible fingerprints cannot travel. Here we see the possibility for rendering othered bodies illegible to the biometric scanner in ways that may have serious implications for people's freedom and mobility. Thus in addition to lengthier and more intrusive forms of surveillance, biometric technologies may also deprive certain communities of their human right to mobility.

In her groundbreaking book on AIDS, the feminist science studies theorist Paula Treichler (1999:162) notes that the majority of researchers doing cultural studies research on the AIDS epidemic "have at least a rudimentary theoretical grasp of virology and immunology. The same cannot be said of most scientists' grasp of social and cultural theory." This holds true for biometric scientists; their ignorance of even the most fundamental tenets of theories of race and racism is deeply problematic, and is made evident in their claims that biometrics might usefully help scientists to reexamine the credibility of physiognomy (Golomb, Lawrence, and Sejnowski 1991) or the suggestion that anthropometry is the basis of modern anthropology (Hua-ming, Zhou, and Geng 2004). This lack of familiarity with contemporary knowledge regarding the socially constructed nature of identity raises questions regarding the knowledge base upon which these scientific practices depend. Clearly the impact of locational differences on biometric functioning, such as Japanese scanners that have trouble recognizing non-Japanese faces, tells us that these technologies are not created in a laboratory environment free of cultural influence. As these scientists label the images according to their understandings of their own biological race and gender identities, preconceptions about gender and race are codified into the biometric scanners from the beginning. Thus assumptions that women are more likely to have long hair and men are more likely to wear ties or that the "Mongolian race" has a certain type of eyelid are reified and then programmed into what David Lyon (2003) calls a "technology of social sorting." Biometric scientists base their studies on dangerous, racist understandings of identity from the 1930s and earlier and ignore the past thirty years of research that has definitively disproved any biological theory of race.

Racial Profiling

The terrorist has a different look, a different face. . . . If I see someone (who) comes in that's got a diaper on his head and a fan belt wrapped around the diaper on his head, that guy needs to be pulled over.

CONGRESSMAN JOHN COOKSEY OF LOUISIANA, CITED IN REEM BAHDI, "NO EXIT: RACIAL PROFILING AND CANADA'S WAR AGAINST TERRORISM"

Biometric technologies have had an interesting role to play in concerns about racial profiling after 9/11. In his article analyzing the increase in racial profiling following Canada's war on terrorism, the legal theorist

Reem Badhi (2003) notes that the racial profiling of Arabs and Muslims was promoted as being in the interest of national security. Certainly the epidemic of racist incidents targeting individuals who are imagined by various security, immigration, and other institutional personnel to bear the signifiers of Arab and Muslim identities have made it clear that racial profiling is an integral part of the national security strategy of both Canada and the United States, revealing the discursive shift from "Is racial profiling occurring?" to "Should it continue to be condemned?" One example of this is the roundtable facilitated by the editors of the *New York Times* in which they posed the question "Will profiling make a difference?" to a wide range of security experts, engineers, policymakers, and FBI agents (2010). One might ask if this question has already been answered given explicit decisions in the United States to racially profile its citizens. For example, the National Security Entry-Exit Registration System requires men (both visitors and immigrants) from a list of countries born on or before November 15, 1986, to undergo special registration procedures, including the collection of their biometric information. The countries on this list include Afghanistan, Algeria, Bahrain, Bangladesh, Egypt, Eritrea, Indonesia, Iran, Iraq, Jordan, Kuwait, Lebanon, Libya, Morocco, North Korea, Oman, Pakistan, Qatar, Saudi Arabia, Somalia, Sudan, Syria, Tunisia, the United Arab Emirates, and Yemen. Over forty local, national, and community organizations in the U.S., from human rights groups to faith-based coalitions and immigrant advocacy groups, called for an audit of this program in 2009 because it has increased the intimidation and harassment of Arab and Muslim communities.

It is no accident that biometrics play a significant role in racial profiling after 9/11. Held up by industry and government proponents as objective and free from the forms of systemic discrimination that plague real life, biometric technologies are represented as particularly useful because they will help to eliminate racial profiling. Biometric technologies, which claim to eliminate racial profiling through mechanical objectivity, simultaneously are explicitly based upon assumptions that categorize individuals into groups based on phenotypical markers of racialization and used in the implementation of programs specifically aimed at racial profiling. Moreover biometric technologies result in new forms of technologized racial profiling as people may not be allowed to travel as a result of how the scanners work. Clearly biometrics allow the state to both implicitly and explicitly engage in racial profiling while using the rhetoric of technological neutrality and mechanical objectivity to obscure this fact.

Queer Renderings of the Body

In the studies described earlier old assumptions about the ways that sex may be read off the body, from hair length to clothing, are assertions about the absolute nature of the divide between men and women. Those bodies that do not fit the inflexible criteria defined for men and women regularly are discounted, as we saw in the Yang, Li, and Haizhou (2006) study, which deemed any face that had a "strange" hairstyle to be unclassifiable. These studies are reminiscent of earlier research attempts to read sexual variance and sexual deviance off the body, as in the case of the medical doctors who attempted to identify lesbianism by measuring vulva size and nipple length (Terry 1990). With respect to biometrics, we might imagine discounted bodies to regularly include those of butch women and femme men as well as transgendered persons, as the photos of those who failed to be correctly identified by Yang et al.'s system suggest. Although Golomb, Lawrence, and Sejnowski (1991) make a reference to "sex outliers" in their foundational study of computerized sex classification, this remains a deeply problematic category. In general people who cannot easily be categorized as either men or women are interpreted as biometric system failures. Showing a naïve belief in the assumption that low-cut blouses and long hair can be used to distinguish women from men with beards and ties, the scientists conducting these studies rely on narrow definitions of sex and gender in which both are collapsed, and in which those who do not easily fit these categories are erased altogether. As Paula Treichler commented, these scientists appear to be immune to even the most basic tenets of fashion; even one season of *Project Runway* might make a world of difference to their attempts at classification.

Although biological understandings of race and gender have long been analyzed by cultural theorists and scientists, their findings have largely failed to make it into the labs of scientists designing biometric systems. Biometrically producing reified racialized and gendered identities can have severe material ramifications. Most at risk from having their race, sex, and gender identities biometrically codified are those who refuse neat categorizations as well as those whose bodies the state believes to be a threat.

What are the implications of biometric techniques of verification for vulnerable bodies? What happens when biometrics fail? Given that the studies I've discussed assume neat bifurcations of gender into the mutually exclusive categories of male and female, the transsexual and transgendered community in particular faces significant risks from biometric forms of identification. For example, the U.S. Real ID Act of 2005 calls for the development of a national database of information and the creation of national identification cards equipped with chips that would carry both biometric data and information about biological sex. The Act poses grave risks for transgendered people living in the United States and was identified as one of the most pressing issues facing transgendered people today by Mara Keisling (2007), the executive director of the National Center for Transgender Equality. Any form of identification that contains biometric information could easily endanger trans folks if it was used to link them to a history of name changes or spotlight discrepancies in legal names and gender markers. In particular it could publicly identify people by making visible connections to outdated gender designations. In this way the representation of biometrics as able to stain identity onto the body—especially given the essentialized notions of gender upon which they are based—makes this community vulnerable to the prying eyes of the state. Mobilization also is occurring in the U.K. to address difficulties created by British plans for biometric IDs and for a national identification card. It remains unclear whether British ID cards will compare a person's biometric profile with his or her "birth gender" or "chosen gender" (Oates 2006).

Summary

Scientific studies that rely upon biological understandings of gender and race represent some of the most egregious failures of biometric technologies. Any technology that takes as its premise the assumption that bodies are stable entities that can be reliably quantified is problematic. Relying upon erroneous biological understandings of race and gender in the development of biometric technologies has a number of ramifications, from the marginalization of transgendered bodies to facilitating forms of mechanized racial profiling. Like other identification technologies before them, biometric technologies are deployed in ways that remind us of other racist regimes premised on similar strategies of racialized and gendered classification.

Using technology to tell us "truths" about the body never reveals the stable narratives we are hoping for. Biometric technologies cannot be counted on effectively and definitively to identify *any* bodies. However, as these technologies are specifically deployed to identify suspect bodies, the impact of technological failure manifests itself most consistently in othered communities. Representing these new identification technologies as able to circumvent cultural assumptions and subjective human judgment does not make it so. Rather biometric technologies are always already inflected by the cultural context in which they were produced. Biometrics are marketed as able to eliminate systemic forms of discrimination at the same time as they are produced in a context marked by the persistence of problematic assumptions about difference. That is, the rhetoric of scientific neutrality masks their racist, sexist, homophobic, ableist, and classist practices. Given the context for which they were developed, it is unsurprising that biometric technologies are imagined as able definitively to identify suspect bodies. Nor is it surprising, given cultural assumptions about othered bodies, that these assumptions are both explicitly and implicitly coded into the technologies. Biometrics fail precisely at the task which they have been set: to read the body perfectly, and in doing so tattoo permanent identities onto deviant bodies. Thus biometrics fail most often and most spectacularly at the very objective they are marketed as able to accomplish. Race and gender identities are not nearly as invisible to new identification technologies as is claimed. The technological fallibility of biometrics manifests itself practically in their disproportionate failure at the intersection of racialized, queered, gendered, classed, and disabled bodies, as they represent the latest attempt to definitively tie identity to the body. Rather than telling stories of mechanical objectivity, race neutrality, and the guaranteed detection of formerly invisible bodies, biometric technologies continue to tell stories heavily inflected by the intersection of bodily identities.

two I-TECH AND THE BEGINNINGS OF BIOMETRICS

Biometrics and policing are not strangers to each other.

ANN CAVOUKIAN, PRIVACY COMMISSIONER OF ONTARIO, *BIOMETRICS AND POLICING: COMMENTS FROM A PRIVACY PERSPECTIVE*

The demonstration of the new iris scanner was identical to technical demos staged at biometric industry conferences aimed at proving how efficiently their products work. For Charles Coney, however, there was a significant difference: he could not refuse to be tested. Coney is a prisoner at Iowa's Story County Jail. He is one of those compelled to use this new technology, bankrolled by the government and rolled out across the United States.

As with any origin story, the history of biometric technologies does not have a tidy timeline. The biometric industry began as a series of fragmented endeavors loosely organized around a generalized interest in access control. Although biometrics made their debut in a number of different spaces, from banks to the workplace, the prison industrial complex is the first program to broadly adopt these new identification technologies.[1] Previous research helpfully grounds the growth of the biometric industry in the financial sector (Gates 2004; van der Ploeg 2005). In this chapter I build on this research and additionally track the beginnings of biometrics to the first major government-funded program to use biometric identification technologies. I am particularly interested in the rise of the prison industrial complex as a site of surveillance that disproportionately targets and criminalizes particular communities in ways that are tied to their gender, race, class, and sexual identities.

Biometric companies capitalized on rising rates of incarceration in order to create a profitable market for their products. Using narratives of corporeal fetishism to represent prisoners' bodies as strands of code capable of yielding their own product inventory numbers in ways imagined to facilitate the management of prison life, state projects use these new identification technologies to create new forms of knowledge about othered bodies. Use and exchange values are wrested from prisoners, whose criminalized bodies are treated as the raw material of capitalism, the fuel necessary to feed the ever-expanding prison economy.[2] Biometric science organizes bodies—bodies imprisoned according to gender, race, disability, class, and sexual identities—in the service of capitalism. Moreover biometric technologies are key to what are broadly termed systems of "knowledge engineering" (Lynch and Rodgers n.d.), processes meant to replace human decision makers with computers. As a result the inclusion of biometric technologies in the prison industrial complex has significant implications for prisoners, including the removal of human witnesses to their suffering and the increased surveillance of prison life. I begin by describing the history of the invention and development of biometric technologies, the disparate companies working independently on these new identification technologies, and some of the early vocabulary developed to describe the science of identification. I then show that early industry focus was most successful for those companies interested in marketing their products to the prison system. I examine the growth of biometric technologies against the backdrop of the expansion of the prison industrial complex. Analyzing the ways that the intersection of poor people, people of color, and women were produced as the primary targets of biometric identification, I show that biometrics were used to police an increasingly broad range of criminalized individuals, from those warehoused in the U.S. penal system to newcomers to the country. I end this chapter by examining some of the implications of adding biometric technologies to the prison system, including the ramification of using prisoners as a test bed for new technologies as well as the consequences of codifying flawed science.

The History of Biometric Technologies

From bertillonage to portrait painting to ink fingerprinting, analog biometric technologies that measure the body for the purposes of identification have a long-standing history (C. M. Robertson 2004; Gates 2004;

Lalvani 1996; Gould 1996). However, my primary concern is the invention of digital biometric technologies, including the new possibilities that networking digital technologies provide for information sharing. As companies with no obvious customers floundered in their attempts to develop a product that could appeal to mass markets, a number of different technologies were researched and tested separately.

Although early biometric research focused primarily on digital fingerprinting, one of the first biometric technologies on the market was a hand geometry reader developed during the 1960s.[3] The company Identimation initially sold biometric hand scanners to a Wall Street investment firm to be used as time clocks for the firm's employees (Sidlauskas and Tamer 2008). Hundreds of these technologies were sold in the late 1970s, and the Identimat remained available until 1987 (Sidlauskas and Tamer 2008). In 1979 Wackenhut bought Identimation. Wackenhut is the largest private security company in the United States and a key player in the privatization of prisons both in the United States and worldwide (Sidlauskas and Tamer 2008). Biometric retinal scanning began shortly after hand geometry, making its debut in the 1970s. In 1975 Robert "Buzz" Hill founded the company EyeDentify. Hill's technology was patented in 1978, and the first working prototype became available in 1981 (Hill 1999). Early retinal scanners were expensive, costing up to $60,000; by 1988 EyeDentify's scanner sold for approximately $7,000 (M. Browne 1988). Although biometric retinal scanners were first used for access control in high-security buildings such as the Pentagon, once they had come down in price prisons bought them for the purposes of prisoner identification (Lichanska n.d.). Not to be confused with biometric retinal scanning, some of the earliest biometric research concentrated on iris scanning. Frank Burch, an ophthalmologist, proposed the idea of using iris patterns for personal ID in 1932 (National Center for State Courts 2002). However, the idea did not take off until the intervention of John Daugman, a scientist who created algorithms for iris recognition which he patented in 1994, the year that biometric iris scanners were first used in the prison system. Daugman's algorithms are the basis for all current iris recognition systems (National Science and Technology Council Subcommittee on Biometrics 2006).

Early biometric developments centered on generalized attempts at access control. That is, in attempting to pitch their products to a broad market, access control was one of the key ways of speaking about the utility of biometric technologies. For example, early retinal scanners were used to control employees coming into work (Reuters 1986). Biometric prototypes also were used for access control in banks and other secured spaces, such as U.S. Navy facilities (Johnson 1984). Interestingly the term *access control* itself grew out of postwar telecommunications.[4] It was listed in a glossary of telecommunications terms in 1949. Given that developments in security and computing technologies generated new language and, as a result, new markets in access control, prisons seemed a natural fit for an industry preoccupied with securing spaces. Obviously the prison is a site understood to require significant access control. In addition prisoners themselves represented "acres of skin" to a biometric industry in its infancy,[5] and one requiring a broad population upon which to test its products. Fingerprinting remains the technology most closely associated with law enforcement and the prison system, and thus biometric fingerprinting was a primary focus of the early industry.

Biometric Fingerprinting and the Prison Industrial Complex

During the 1950s the FBI began using punch cards to store information about fingerprints; thus the origins of biometric fingerprinting date back to the early days of computing (S. A. Cole 2004). It is notable that the FBI's search-and-retrieve fingerprint system relied upon information about identity from the outset: "By assigning a punch card encoded with information such as the individual's gender, age, [and] race . . . to each fingerprint card, operators could then do a card sort searching for certain parameters, and operators could then retrieve the matching fingerprint cards" (17). Here we see that assumptions about the fixed nature of race and gender identities were central to the classification and surveillance of individuals involved in the prison industrial complex from the beginning.

In the 1960s the FBI sponsored research at the National Bureau of Standards on the potential of further automating fingerprint identification and searches. A major research paper on the possibility of this project was published in *Nature* by Mitchell Trauring in 1963. Tauring's paper had a significant impact on the development of early biometric technologies, and is an important and heavily referenced part of biometric history. His article provided the basis for the continued development of biometric fingerprinting and is one of the first papers published in the field of automated biometric identification (Wayman 2004). The first patent awarded for biometric fingerprinting technology was issued to IBM, awkwardly titled "Personal Security System Having Personally Carried Card with Fingerprint Identification" (Wayman 2004). Attempts were made in the late 1960s to use holographic imaging for digital fingerprinting, but the technology proved too expensive (S. A. Cole 2004). New techniques facilitating the enhancement of images developed for NASA during the late 1960s and early 1970s significantly improved the process of digitizing fingerprinting (Cherry and Inwinkelried 2006). The first biometric fingerprinting system was tested in 1970 by what became the National Institute of Standards and Technology, then known as the U.S. National Bureau of Standards (Wegstein 1970). Automated fingerprint identification systems (AFIS) also were central to biometric fingerprinting development.[6] This technique was debuted in 1972 by the FBI (S. A. Cole 2004). In the mid-1970s a number of biometric companies—including Fingermatrix (1999) in the United States, which has been in business since 1976, NEC (Japan), and Morpho (France)—began marketing different versions of AFIS digital fingerprinting technologies to law enforcement agencies (S. A. Cole 2001:253–54).[7] Both NEC and Morpho had made successful inroads into U.S. markets by 1985 (S. A. Cole 2001). Although the FBI began using automated search techniques for fingerprints in 1979, these did not become standard until 1983 (S. A. Cole 2004). The prison systems' growing state-backed budgets provided an ideal location for biometric beginnings, as "procurement of an AFIS was . . . in the range of $2 million to $10 million, and the conversion of the database—scanning fingerprint cards into digital form—could cost hundreds of thousands of dollars" (S. A. Cole 2001:253). In 1983 the FBI reported that it had digitized its entire fingerprint file, right back to 1928 (S. A. Cole 2001). By 1986 police bureaus had embraced digital AFIS, giving rise to "a frenzy of Request for Procurements" (253).

Given that AFIS shifted the standards for fingerprinting identification in addition to dramatically increasing search capabilities and the speed with which searches could be conducted, the fingerprinting historian Simon Cole notes that AFIS has justly been placed alongside DNA typing as "one of the two most important advances in forensic science in the 20th century" (2004:19). Biometric fingerprinting technologies made it "technically feasible to routinely check the fingerprints of everyone who comes into contact with the police," profoundly shifting identification practices in the United States (S. A. Cole 2001:257). Thus biometric technologies are part of a larger scientific trend toward corporeal fetishism; that is, these new technologies are used to codify bodies in ways that claim individual identities to be both knowable and searchable, without reference to the ways that biometric codifications of the body are themselves troping devices that draw on existing cultural assumptions that individuals can be broken down into their component parts and sold as commodities. In representing complex, relational bodily identities as discrete, reified things, biometric science eased the passage of individual identities to market in the prison industrial complex.

One of the significant expansions in biometric fingerprinting systems occurred in 1989, when Congress provided initial funding to what was then Immigration and Naturalization Services (INS) to develop an automated biometric fingerprint system for newcomers to the United States (U.S. Department of Justice 2001).[8] The original goal of this program was to identify individuals who were repeatedly apprehended attempting to enter the United States without being inspected as well as those individuals who previously had been deported, were suspected of criminalized activity, or had outstanding warrants for their arrest (U.S. Department of Justice 2001). The expansion of biometric fingerprinting to newcomers to the U.S. fits in more broadly with an agenda aimed at the increasing criminalization of immigration, one that furthers xenophobic policies while generating corporate profits by expanding the prison industrial complex through building new immigration and refugee detention centers. Between 1991 and 1993 INS piloted its first AFIS system at the U.S. border with Mexico in the San Diego Border Patrol Center (U.S. Department of Justice 2004). Fearing delays for applicants to INS as well as coordination issues with the FBI, INS initially abandoned this pilot project and instead continued with its own IDENT system rather than the FBI's AFIS system (U.S. Department of Justice 2001). IDENT made its debut in 1994. Congress then mandated that the FBI's AFIS fingerprint system be integrated with the IDENT immigrant identification system before they would give the INS any additional funding to develop IDENT (U.S. Department of Justice 2000). In the coordination between INS and the FBI, biometric technologies were part of the increase in state systems of surveillance, rendering immigrant and refugee bodies suspect and extending technologies formerly limited to law enforcement to immigration in a move sometimes termed "crimmigration" (T. Barry 2010).

In addition to technological challenges for the INS and the FBI, standards between law enforcement agencies were incompatible. Although few law enforcement agencies were without a repository of AFIS fingerprints by the 1990s, in 1992 a survey found that twenty-nine local law enforcement agencies used Printrak in their storage of fingerprints, twenty-six used the NEC system, and seven used the Morpho system. FBI prints were stored in another manner altogether, in a computerized system known as FINDER (S. A. Cole 2001:253–54). Because the various AFIS databases were not linked, agencies were unable to share information or conduct national searches for a particular set of fingerprints. In 1994 the FBI sponsored a competition for a technological solution that would allow AFIS repositories to be connected to form a giant fingerprint database, linking hundreds of AFIS databases, making it a searchable, national fingerprint database. The FBI awarded Lockheed Martin the contract to build the IAFIS database (National Science and Technology Council Subcommittee on Biometrics 2006). IAFIS's homepage describes the system as a

> national fingerprint and criminal history system maintained by the Federal Bureau of Investigation (FBI), Criminal Justice Information Services (CJIS) Division. The IAFIS provides automated fingerprint search capabilities, latent searching capability, electronic image storage, and electronic exchange of fingerprints and responses, 24 hours a day, 365 days a year. As a result of submitting fingerprints electronically, agencies receive electronic responses to criminal ten-print fingerprint submissions within two hours and within 24 hours for civil fingerprint submissions. (IAFIS 2008).

IAFIS is now undergoing expansion into a system called, in true *Star Trek* fashion, Next Generation Identification (NGI). In 2008 Lockheed Martin won a $1 billion, ten-year contract to expand IAFIS into NGI for the U.S. government. NGI is "the world's largest biometric database" (Lipowicz and Bain 2008). Although "fingerprints will still be the big player" (Lipowicz and Bain 2008), the Next Generation program will also include biometric iris information, palm print information, and tattoo mapping, and is expected to include a greater range of biometric information down the line.

The 1990s also mark the expansion of biometric technologies to other programs, mirroring the growth of state surveillance systems, including their application to the U.S. welfare system. But the history of biometrics is not a neat chronology; the 1990s also is when biometric technologies began to expand from strictly access control and law enforcement applications to their introduction in prisons themselves. Although the first prison to include biometric retinal scanners to identify prisoners opened in Utah in 1988 (M. Browne 1988), prison use of biometric identification technologies occurred largely after 1990. In that year the Cook County Prison adopted biometric retinal scanning in order to track its prisoners. By 1995 it claimed to have identified 350,000 prisoners using retinal scanning technology (Ritter 1995). In 1994 Pennsylvania's Lancaster County Prison became the first to use biometric iris scanners for prisoner identification (QuestBiometrics 2005). In the late 1990s adding biometric technologies to prisons became a booming business. Identix biometric technologies were increasingly adopted in California's prison system between 1997 and 1999 (Beiser 1999). Between 1999 and 2002 biometric systems were installed at prisons in Florida and Massachusetts, expanding to include facial recognition technologies ("State Lines: Digital Prison Stripes" 2004). In 2000 the National Institute of Justice (the research arm of the Department of Justice) and the Department of Defense embarked on a research study comparing the use of different biometric technologies in prisons (Miles and Cohn 2006). In 2005 the International Biometric Group was awarded two grants totaling $900,000 from the National Institute of Justice to study prison applications of biometrics (International Biometric Group n.d.). Following 9/11 the biometric industry turned to bigger fish, including the multibillion-dollar contracts being offered by the newly formed Department of Homeland Security.

The Expansion of the Prison Industrial Complex

Biometric technologies were developed for the prison system at a time when the system itself was undergoing tremendous expansion. The massive growth in prisons of the surveillance, policing, and criminalization of marginalized communities is well documented (A. Y. Davis 1981, 1998, 2003, 2005; Garland 2001a, 2001b; Sudbury 2004; Gilmore 2007). With a dramatic shift from rehabilitation to punishment in the 1970s (Gilmore 2007) and in part as a result of "three strikes" laws, truth in sentencing initiatives, the war on drugs (Cole and Mobley 2005; D. E. Roberts 2001), the shift from a welfare state to a neoliberal state, resulting in dramatic cutbacks to social programs (Wacquant 2009; A. Y. Davis 2003), and the surveillance and criminalization of immigration, the number of people in prison has exploded in the past thirty years. Prison populations have risen by a factor of ten, from 200,000 in the late 1960s to more than 2 million by 2000 (A. Y. Davis 2003). Today the United States has more people behind bars than any other country (Hartney 2006), an astonishing 25 percent of the world's prisoners, although the country has only 5 percent of the world's population (Critical Resistance 2000). The role of the prison industrial complex as an engine of inequality through the disproportionate incarceration of both poor people and people of color is widely noted; for example, 1 million African Americans are behind bars (Sudbury 2004), and despite studies finding little difference between drug use among people of color and white people (J. Webb 2009), over two-thirds of those in prison or jail for drugs are people of color (S. A. Cole 2007).

Nor can we disregard gender in theorizing inequality in the prison system. In keeping with worldwide trends in which poor racialized women and men with mental health disabilities are the fastest growing groups to be incarcerated (CAEFS 2004), African American women are now the fastest growing prison population, having outpaced African American men (A. Y. Davis 2003).[9] The incarceration of poor women (and poor people generally) must be placed alongside the dismantling of the welfare state. The elimination of welfare programs like Aid to Families with Dependent Children causes women to seek out criminalized forms of employment so that they can afford food and housing. Sexual, emotional, and physical violence also propel women into the prison system (CAEFS 2004). As women flee abusive situations, the lack of a social safety net means they must turn to criminalized forms of labor such as sex work and the drug trade in order to meet their subsistence needs (Sudbury 2005; A. Y. Davis 2003). (It is important to note that, in addition to the fact that a huge proportion of women who end up in prison are survivors of sexual violence, the prison itself is a source of sexual violence, both through the rape of prisoners by guards and the state-sponsored sexual assault of the strip search; Sudbury 2005; A. Y. Davis 2003).[10]

Homophobia, transphobia, and heterosexism and their intersection with sexism and racism render LGBTQ people vulnerable to the prison industrial complex. As Beth Richie's (2005) study of young black lesbians shows, the relationship between homophobia and sexual harassment places queer women at increased risk of violence, sometimes leading to their engagement in illegal activities. Queer people have historically been and continue to be disproportionately caught up in the prison system, as queered sexualities are criminalized by the state (Kunzel 2008). Homophobia also continues to play a significant role in sentencing and in the space of the prison. Queers are subject to disproportionate violence while incarcerated. For both queer men and queer women who were kept in separate prison wings, these spaces often resulted in worse treatment, including no programming and continual confinement with fewer activities than other prisoners (Kunzel 2008; Goldstein 2001). The role of homophobia in sentencing is understudied, but preliminary research indicates the ways that queer identities can result in more severe punishment. For example, it is well documented that homophobic judges and juries regularly assume that same-sex prisons are utopias for queer men and women and thus have no difficulty sentencing them to confinement (Kunzel 2008; Goldstein 2001). And research indicates that homophobia may have an impact on inciting judges and juries to condemn queer individuals to death (Goldstein 2001). For example, in the case of Wanda Jean Allen, a black lesbian accused of murdering her partner, part of the evidence that appears to have fueled the jury's decision to execute Allen was that she was a butch woman who "wore the pants in the family" and preferred the male spelling of her middle name (Gene). Evidence of Allen's butchness was used to question any suggestion that she and her partner were both abusive to each other, and instead was used to prove that she had used her masculine power to overwhelm her partner, evidence that played a role in the jury's decision to execute her, the first woman to be put to death in the state of Oklahoma. It is additionally notable that Allen did not have any funds to hire a lawyer, and the lawyer assigned to her case was paid only $800 (Goldstein 2001).

Disability also makes people more vulnerable to criminalization: "The U.S. Department of Justice estimates that 16% of the adult inmates in American prisons and jails—which means more than 350,000 of those locked up—suffer from mental illness, and the percentage in juvenile custody is even higher. Our correctional institutions are also heavily populated by the 'criminally ill,' including inmates who suffer from HIV/AIDS, tuberculosis, and hepatitis" (J. Webb 2009). Women with disabilities are one of the fastest growing groups to be incarcerated. As the prison system replaces state-funded efforts to address mental and physical health, people living with disabilities are increasingly caught up in the prison industrial complex.

The expansion of the prison industrial complex must be situated against the rise of global capitalism and the profits made from increased surveillance. As the U.S. spearheads an international turn toward neoliberal policies, the welfare state is being dismantled (Wacquant 2009), and prison building has risen to take its place (Sudbury 2005; A. Y. Davis 2003; Gilmore 2007; J. James 2007). Prison populations have risen alongside the decline of New Deal programs, including those supporting education, health, and welfare, meaning that members of many communities of color—including Native Americans, Latinos, and African Americans—have a greater chance of going to prison than of getting an education (Wacquant 2009; A. Y. Davis 2003). This has led some prison scholars to assert that "short of major wars, mass incarceration has been the most thoroughly implemented government social program of our time" (Currie 1998:10). As Dorothy Roberts (2001:91) explains, "The prison became our employment policy, our drug policy, our mental health policy, in the vacuum left by the absence of more constructive efforts."

The prison economy is big business. Warehousing 2 million people behind bars has provided companies like the Corrections Corporation of America and Wackenhut with unprecedented opportunities for profit (Myser 2007). Having one in every thirty-one adults in the U.S. caught up in the prison industrial complex—whether incarcerated or under surveillance following or in place of literal confinement—is extremely costly to taxpayers (J. Webb 2009). Local, state, and federal spending on corrections adds up to about $68 billion a year in the U.S. alone (J. Webb 2009). The drive for profits has led to many instances of abuse. Guards working for the Corrections Corporation of America complained "that they were encouraged to write up inmates for the most minor infractions and place them in segregation, which takes away points they've established for good behavior. It also adds a full 30 days to their sentences, which can help make about $1,000 for the prison in pure profit" (Frank 2009). One prison run by Corrections Corporation of America was described as a "hotel that's always 100% occupancy . . . and booked to the end of the century" (Welch and Turner 2007). In 2009 two judges pleaded guilty to charges that they took millions of dollars in kickbacks from a private prison company in return for sending juveniles to the company's privately run juvenile detention facilities (Urbina 2009; Freed 2009). The judges were discovered partly as a result of their exceedingly harsh approach to juvenile punishment:

> Mr. Conahan's alleged partner in the scheme, Judge Mark Ciavarella Jr., reportedly sent kids to the private detention centers when probation officers didn't think it was a good idea; he sent kids there when their

crimes were nonviolent; he sent kids there when their crimes were insignificant. It was as though he was determined to keep those private prisons filled with children at all times. According to news stories, offenses as small as swiping a jar of nutmeg or throwing a piece of steak at an adult were enough to merit a trip to the hoosegow.

> Over the years Mr. Ciavarella racked up a truly awesome score: He sent kids to detention instead of other options at twice the state average, according to the New York Times. He tried a prodigious number of cases in which the accused child had no lawyer—here, says the Times, the judge's numbers were fully 10 times the state average. And he did it fast, sometimes rendering a verdict "in the neighborhood of a minute-and-a-half to three minutes," according to the judge tasked with reconsidering Mr. Ciavarella's work. (Frank 2009)

Making money by sending children to jail is just one of the terrible consequences of running prisons for profit.

Biometric companies too profited tremendously from the expansion of the prison system. Biometric technologies are enormously expensive. In the early 1990s Los Angeles law enforcement spent $15 million on mobile biometric fingerprint scanners (Adelson 1992). State police departments across the United States spent hundreds of millions of dollars installing live-scan fingerprint readers for local agencies. Valued at more than $30 million dollars each, live-scan readers and matching computers were installed in thirty-six states by 1992, costing a total of $1.1 billion (Adelson 1992). High-tech companies hoping to develop their technological products for the prison industry have valued sales of new technology at $1.4 billion (Business Editors and Technology Writers 2001). Clearly the addition of biometric technologies to the prison system must be located within the growth and privatization of the prison industrial complex itself.

Implications of Adding Biometrics to the Prison Industrial Complex

Given the astronomical start-up costs of biometrics, the industry had to find significant funding to get its technologies launched. In applying its technologies to the prison system, the industry found the necessary research and development funds from a combination of government subsidies, grants, and private-public partnerships. It is the billions of dollars of funding for systems like AFIS, IDENT, and other fledgling biometric surveillance systems that enabled this technology to get off the ground.

PRISONERS AS TEST BEDS FOR BIOMETRIC TECHNOLOGIES

The prison industrial complex offered another resource to a fledgling industry testing out a new product: prisoners' bodies are valuable commodities to biometric companies, providing the industry with a captive test population for assessing the efficacy of these new identification technologies. While the banking industry might be cautious about introducing biometric technologies to their clients for fear of scaring off potential customers, prisons do not have the same restrictions. In this way biometric technologies are key to the intensification of particular forms of monitoring, as they provide a way for companies selling surveillance technologies to try out products that would be rejected by the general public. As Calum Bunney, editor of the magazine *Biometric Technology Today*, based in the U.K., asserted:

> In the outside world, having your retina or iris scanned can be an inconvenience. Not so in the coercive environment of the clink, where biometric technology is used to great effect. A bank will obviously not want to put its customers off by falsely rejecting them when they want to access their account via a biometric ATM. . . . A prison, on the other hand, has a captive audience, and can make the user of a biometric device perform the actions required for passing as many times as it likes until it is satisfied. Who cares if the customer doesn't like it? (Glave 1997)

The biometric industry used the prison industrial complex as a giant laboratory where they could refine their technologies without fear of reprisal. The rights of prisoners to refuse biometric technologies were considered negligible:

> Larry Cothran is pretty frank about the degree to which civil rights and privacy issues figure in his work. They don't. "When you come on to a state property, you give up most of your rights," Cothran says.[11] As executive officer of the Technology Transfer Program with the California Department of Corrections, Cothran evaluates and recommends systems that are expressly designed to watch, detect, secure, and contain some 155,500 inmates in 32 prisons across California. (Glave 1997)

Testing biometric technologies in places where they can't be refused has a number of implications. For example, biometric retinal scanning is not in widespread use, partly because retinal scans are very difficult and time-consuming to take, and people dislike having their eyes so close to the scanner. Another reason is that biometric retinal scans can reveal significant amounts of health information, including pregnancy, HIV status, diabetes, and other information that might have a significant impact on the person's health privacy (Shaikh and Dimitriadis 2008; Kiruba 2005). Although this presents an obstacle for their development for other applications, because prisoners cannot decline to be scanned means these technologies can continue to be refined and find a market in the prison system.[12]

PRISONERS AS HUMAN INVENTORY

"In risk society, policing is not just a matter of repressive, punitive, deterrent measures to control those who are morally wrong. It is also a matter of surveillance, of producing knowledge of populations that is useful for administering them" (Ericson and Haggerty 1997:41). Biometric technologies are used to produce new forms of knowledge about prison populations. Producing every incarcerated body as a strand of binary code paves the way for understanding prisoners as human inventory. It is a form of corporeal fetishism that reduces complex human bodies to products used to generate value for the prison industrial complex. Connections between merchandise and prisoners are often made explicit:

> In a small-town grocery store, a routine inventory is under way. Clerks move quickly among the shelves of canned goods and boxes of pasta, holding scanners in their hands, passing them over bar codes and flashing information back to the store's central database.
>
> Down the road, in the state correctional facility, officers move among inmates "counting heads"—not just once, but several times during the course of the day. Another type of routine "inventory" is being conducted, but this one consumes much more time and resources. Soon, however, correctional officers may have access to improved technology that makes counting inmates go as quickly, smoothly and accurately as other inventory counts. (Lewis 2003)

There are implications to using biometric technologies to count heads in ways that reduce the bodies of criminalized individuals to biometric inventory. For example, counting heads provides an opportunity for guards to determine if a prisoner is ill, injured, or otherwise in distress. It also provides inmates with a set of eyes to bear witness, including to forms of prisoner resistance or protest. How effective would a hunger strike be if a biometric scanner were the only bystander? The addition of biometric technologies to prisons also has consequences for prisoners' health and well-being, as well as for the means of protest available to them. Biometric representations of the body as a simple series of ones and zeroes reinforce existing ideologies in which prisoners are reduced to numbers that need to be counted instead of being understood as human beings to whom we need to attend.

BIOMETRIC TECHNOLOGIES AND KNOWLEDGE ENGINEERING

In a short paper detailing the benefits of adding biometrics to law enforcement, Clive Reedman (n.d.), a biometrics industry representative, describes one of the central advantages of these new identification technologies as "automation." Reedman explains that, if compelled to identify a large population for hours on end, he "will become tired, bored, or distracted: I will need to take a break and eventually sleep. My concentration will fail and I will make mistakes." Unlike humans, biometric scanners never get tired, are never lazy, nor do they commit errors. As biometric technologies regularly are represented as able to improve upon human failings, they are easily incorporated into what are broadly termed systems of "knowledge engineering," processes that are meant to replace human decision makers with computers: "Knowledge engineering is an iterative process that allows the expert and knowledge engineer to continue to refine the rules and facts embedded in the expert system until it functions at the level of the expert. Unlike conventional computer programs, expert systems are designed to be readily modifiable to incorporate new, conflicting, or incomplete information" (Lynch and Rodgers n.d.). In this way biometric technologies are imagined to be able to improve upon subjective human experts by providing mechanical objectivity, eliminating the need for human intervention. This rhetoric hides how these new identification technologies both codify and intensify existing cultural assumptions and inequalities.

LABOR, BIOMETRICS, AND SURVEILLANCE

Questions of labor are central to understanding the growth of the prison industrial complex. One of the reasons that privatization can be marketed as cost-effective is because it paves the way for reductions in pay. Rather than being compelled to hire highly trained, unionized prison workers, privatization allows companies to hire personnel and give them little to no training, keep them permanently on part-time work, and thus reduce their wages. As Pat Cannan, a representative with Wackenhut, says:

> "We don't pay a lot of overtime, and maintain a part-time work force." Stock options in the company are given rather than cash, if for no other reason than that it's usually cheaper for these businesses to do so; in 1996, annual earnings for unionized prison staff was about $33,330, while the non-union boys, even with the stock options the companies love to talk about, only made about $24,000—or a whole third less in total income. (Montgomery 2001)

Paying workers less is only one part of the prison industry's strategy for cutting down on expensive labor costs. Another strategy is to replace human personnel with machines. Biometric technologies are central to this strategy. As industry experts reported, biometric technologies eliminate the need for human screening, helping to cut costs.

> Before they discovered Schlage Recognition Systems HandKey, Northern Ireland Prison Services monitored all comings and goings at its prisons by posting guards at all access sites and supplying them with keys. This was an expensive policy, given the guards' high salaries. . . ."We were able to install a pair of HandKeys for a one-time payment of less than one-fourth of what we paid each guard annually," said Michael Pepper, Director of Activities and Services at Maghaberry Prison. "And costs were not the only consideration. HandKeys are not subject to intimidation, for example, nor do they tire easily." (Schlage 2006)

Identifying prisoners biometrically also allows for intensified forms of surveillance, including of prisoner labor. For example, biometric checks can be used to monitor prisoners while they are at work, including ensuring that inmates have their pay docked if they take too many bathroom breaks:

> "Inmates are always playing games," Donlin [corrections program manager at the National Institute of Justice's National Law Enforcement and Corrections Technology Center] says. "They get paid, say, 40 cents a day to work in the work center and they report to work. But then

> they say they have to go to the doctor, they have to go to the psychologist, they spend the whole day running around, then claim they were there the whole time. Using this system for portal control would track their movements and show that they shouldn't be paid for that day." (Lewis 2003)

Given the consequences for prisoners' health, privacy, labor rights, and well-being, the effect of adding biometrics to the prison system goes far beyond technological efficiency.

Summary

The rise of biometric identification technologies was accompanied by industry and governmental assertions that biometric identification produces the truth of identity, a truth that can be neither distorted nor concealed. Yet given the fraught conditions under which biometric identities are wrested from prisoners' bodies, the integration of these technologies in the prison can hardly be regarded as a success. Although parallel developments occurred in welfare as well as in the financial, military, and health sectors, the driving logic behind the expansion of biometric identification is the control and management of marginalized communities. The development of biometric technologies resulted in the intensification of existing processes of criminalization while increasing corporate profits —calling into question industry claims that biometric technologies function with mechanical objectivity. However, although biometrics fail to function in the neutral ways that industry claims, these technologies succeed even where they fail. Making captive populations productive by using biometric technologies to transform bodies into binary code allows vulnerable bodies to be spun into gold.

With the decline of the welfare state and the intensification of state surveillance, the prison industrial complex has stepped in to fill the gap formerly occupied by the educational system and a more expansive social safety net. Biometric companies with a vested interest in expanding their profits and state legislators with conservative agendas represent these new technologies as the panacea for fixing what is broken in the prison system. Thus technological solutions are offered up to solve social problems, and the same vulnerable communities—poor people, queers, women, and people of color—serve as the raw materials upon which to test new technologies. Besides marking a return to nineteenth-century claims about the possibility of reading criminality right off the body, biometrics have ramifications for the state's most vulnerable populations in ways that reinforce existing forms of systemic inequality, including sexism, racism, classism, and homophobia.

Despite the significant human and economic costs of these technologies, technological solutions to the problems of poverty are on the rise. With their eyes firmly fixed on the potential for profits to be made in the prison system, new high-tech companies are attempting to get themselves a piece of the pie by positing ever more draconian technological solutions to social problems. Most recently injecting prisoners with chips containing biometric information is being offered as the ideal solution to the problem of violent offenders, a problem that comes with a $1.4 billion price tag for the company in question (Business Editors and Technology Writers 2001). And the biometric industry continues to seek other clients for its markets.

three CRIMINALIZING POVERTY
Adding Biometrics to Welfare

The people lining up to have their fingerprints taken could easily be confused with prison inmates. Waiting in line to have their prints digitally scanned, they were among the first to be fingerprinted in order to qualify for welfare benefits. Although many resented the association of their need for benefits with criminality, none refused to have their prints taken: "Like Ms. Taylor, Mr. Jones said he resented the fingerprinting. But in the end, both went inside and placed their fingers on the screen. 'If you need assistance, you go along with the program,' Ms. Taylor said. 'That's the bottom line' " (McLarin 1995a).

Trying to identify markets for their products, early companies attempted to sell biometric technologies for a number of purposes, including bank security and employee surveillance. However, profits occurred only once the industry digitized fingerprinting, a technology already associated with the prison system. Industry successes were also tied to the fact that biometric technologies were tested on a population that could not refuse. This burgeoning industry needed to continue to find new markets for its technologies. The next major application of biometrics was to welfare as part of the intensification of the surveillance of those receiving particular forms of aid from the state.

> Because translating biometric technologies to welfare systems is relatively new, industry leaders say it's virtually unexplored terrain with much opportunity to grow. "It's going to happen," said William Rogers, editor of Biometric Digest. . . ."It's not a question of if. It's a

question of when. There's much interest in this technology from the state and local governments, and welfare is key to their interest. That's where the fraud is and the technology is needed." (McCafferty 1997)

From social security to passport numbers, new technologies play a powerful role in the administration of the state and in the production of the identities of state members. Asking why the biometric industry seized upon the U.S. welfare system as the next major client for its products, in this chapter I track the beginnings of biometric fingerprinting to welfare recipients in California, followed by its growth across twelve states. Grounding the development of biometrics in their cultural context, I examine the expansion of these new identification technologies to social service provision against the transformation of the welfare system in the 1990s.

Expanding Biometrics to Welfare

David Banisar, a member of the Legal Council for the Electronic Privacy Information Center, says, "Fingerprints started out strictly as a criminal thing, then they moved into welfare. . . . Technology used for law enforcement purposes has a way of suddenly becoming used for all kinds of things" (Beiser 1999). California was the first state in the union to biometrically fingerprint welfare recipients. The Los Angeles Department of Social Services began collecting biometric information in 1991 using a system called AFIRM, in which the digital fingerprints of new welfare applicants are checked against an existing database of prior claimants with the aim of deterring welfare fraud (Campbell, Alyea, and Dunn 1997). Biometric fingerprinting quickly expanded to other counties in California. In 1994 welfare clients in San Francisco, Alameda, and Contra Costa counties were biometrically fingerprinted. That same year the biometric identification program in Los Angeles expanded to families receiving Aid to Families with Dependent Children (AFDC), making it the first city in the U.S. to biometrically fingerprint parents with children (McLarin 1995a). California imposed mandatory finger imaging for welfare recipients across seven counties by the second year of operation; in 2000 the state proposed linking more than 225 county welfare offices representing more than 6 million welfare recipients (U.S. Social Security Administration, Office of the Inspector General 2000).

It is no accident that California was the first state to apply this new technology to welfare receipt. California occupies a significant place in the U.S. national imaginary and is often where significant changes to national policy begin. For example, California passed Proposition 187 in 1994, which supported heated anti-immigration sentiment that found purchase in other legislation, including welfare reform (it was later found to be unconstitutional). Proposition 187 was aimed at eliminating "public health, welfare and education provisions from undocumented migrants" (Ono and Sloop 2002:3). Following its initial passage in California, the proposition's legislative goals spread across the nation with the passage of the Personal Responsibility and Work Opportunity Reconciliation Act in 1996, Clinton's legislative answer to "ending welfare as we know it." Thus Proposition 187 signified the changes in national welfare policy to come. Moreover California was the first state in 1988 to privatize its welfare program (McGowan and Murphy 1999), a move that facilitated the increased surveillance of the state's welfare recipients through biometric fingerprinting.

New York, Texas, and Connecticut were the next to adopt mandatory biometric identification for welfare recipients. New York added biometric fingerprinting requirements for welfare recipients starting in 1995, enrolling more than one million people in the first three years (U.S. Social Security Administration, Office of the Inspector General 2000). Also in 1995 the Texas state legislature passed a bill permitting biometric fingerprinting technologies to be used in its welfare system. Biometric fingerprinting subsequently expanded to four hundred welfare offices statewide (Nabors 2003). Connecticut's Department of Social Services began biometric fingerprinting systems for welfare recipients receiving AFDC or who were members of general assistance programs in 1996.

Also in 1996 both Illinois ("Illinois Announces AIM System" 1997) and Wisconsin (U.S. Social Security Administration, Office of the Inspector General 2000) expanded the biometric identification of welfare recipients to include retinal scanning. Facial scanning has also been used on welfare recipients (Associated Press 1996). It is not surprising that Wisconsin was among the first states in the nation to expand biometric programs beyond fingerprinting, as that state enacted some of the most dramatic changes to welfare reform under the governorship of Tommy Thompson, who reduced welfare caseloads by 93 percent, from 98,000 people receiving AFDC in 1987 to 6,700 W-2 cash assistance families in 2000 ("Tommy Thompson on Welfare and Poverty" 2008).

As of 2000 biometric programs for identifying welfare recipients were in operation in eight of the fifty states (California, New York, New Jersey, Connecticut, Massachusetts, Illinois, Texas, and Arizona), with programs pending in three additional states (Florida, Pennsylvania, and North Carolina).

Contextualizing the Integration of Biometric Technologies to Welfare Delivery

The biometric industry's profits remained limited in the law enforcement industry. Although biometric technologies were sold in small quantities to banks and for other industry applications, the industry was interested in finding government-funded programs to buy their products, in large part because the U.S. government is the largest potential client of this new identification technology (Chen 2003). The 1990s generated a welfare gold rush in part because of the intensification of the surveillance, policing, and criminalization of welfare recipients—as the privatization of social services proceeded apace—paving the way for the integration of biometric technologies with the provision of welfare benefits. A number of companies, including Digital Biometrics, Printrak, Fingermatrix, and the industry leader, Identix, eagerly jumped on what they described as the welfare bandwagon, gambling that the next big market for biometrics "in state and local governments will be welfare fraud prevention" (McCafferty 1997). As Mike Lyons, product marketing director at Printrak, asserted, "We see welfare fraud as the main thing. The technology is available and very cost-effective to prevent this kind of fraud." The initial integration of biometric technologies to welfare delivery was profitable. Three years after they sold their technologies to welfare programs, two major players in the industry, Identix and Digital Biometrics, "reported their first profitable quarters ever" (Adelson 1994). In fact following the application of their products to welfare, Identix was the first biometric company ever to report a profitable quarter ("Alanco to Acquire TSI, Inc." 2001), revealing the financial security offered by new forms of state surveillance of those unable to refuse to provide their personal data to the state. As of 1998 Identix products were being used to identify welfare recipients in five states (Identix, n.d.). Although these profits did not persist, and revenues in the biometrics industry continued to fluctuate even after its expansion to welfare,[1] the desire for the state as a customer was one of the driving forces behind its expansion to welfare, and this attempt was partially successful.

During the 1990s the biometric industry became interested in many potential markets. From health care to banking, other avenues offered profit potential. Yet welfare remained particularly attractive for a number of reasons; foremost was gaining the state as a client. However, equally important was the political backlash against welfare recipients. The demonization of those receiving assistance from the state made possible the biometric identification of welfare recipients using a technology formerly reserved for the surveillance of those in the prison system. The backlash against welfare additionally was motivated by problematic stereotypes of welfare recipients, the corporate desire to depress wages in order to cut costs, and policies permitting the privatization of welfare.

The inclusion of biometric identification technologies in welfare programs was part of a campaign of sweeping reforms, begun one year before Bill Clinton pledged to end welfare "as we know it" during his election campaign of 1992 (Abramovitz 2000:13). As others have argued (Kohler-Hausmann 2007; Reese 2005; Eubanks 2006; Little 1998), opposition to welfare is not new, nor is the intense surveillance of welfare recipients (Eubanks 2006; Gilliom 2001; Tice 1998; Wacquant 2009). The battle between governmental regulation and the free market dates back to the beginning of the United States (Starr 1982) and has been a constant struggle between the cult of the rugged capitalist individual and debates as to whether, and how much, the state should care for its population. Although there have been ebbs and flows, the 1970s yielded significant cutbacks to the welfare system (Gilliom 2001; Kohler-Hausmann 2007). These cuts began under the Quality Control welfare movement in the United States. Begun in the 1960s, this movement was in full swing by the mid-1970s (Gilliom 2001:29). One of the defining features of the Quality Control movement was the federal attempt to shift the responsibility for welfare to the states, in large part by providing states with incentives for accuracy while imposing penalties for errors in welfare administration. The Quality Control movement was marked by the increased surveillance of welfare recipients, an emphasis on eliminating fraud, and extensive audits of state welfare administrations, all facilitated by the introduction of early computing technologies to the welfare system (Gilliom 2001:30).

Cuts continued with Ronald Reagan's changes to the welfare system in the 1980s. To achieve his goal of welfare reform, Reagan granted states a freer hand in their ability to tailor aid programs, furthering the elimination of federal responsibility for state welfare systems (Rogers-Dillon 2004; Kahn and Kamerman 1998). This strategy was adopted by Clinton in the Personal Responsibility and Work Opportunity Reconciliation Act (PRWORA) of 1996 (Kahn and Kamerman 1998) and signed by most states in 1997 (Abramovitz 2000). PRWORA delegated federal responsibility for welfare to the states, giving them a free hand in the administration of welfare programs and paving the way for privatization. The majority of states turned to corporations (Brophy-Baermann and Bloeser 2008), fueled by their assertions that privatization would make use of business efficiency models, which relied heavily on new technologies to cut costs. Thus the Clinton-era reforms allowed states the flexibility to introduce biometric technologies to welfare.

Significant about PRWORA was that it targeted immigrants and refugees to the United States. PRWORA forbade legal immigrants to the United States from obtaining welfare for their first five years in the country (Reese 2005:3).[2] Following PRWORA, there were significant reductions in legal immigrant use of all benefit programs. This was in part a result of the new cutoffs, but PRWORA additionally had a chilling effect on immigrant applications to welfare programs more generally, as newcomers to the U.S. became concerned that welfare collection would have an adverse impact on their ability to obtain citizenship (Fix and Zimmermann 1998). Between 1994 and 1999 Temporary Assistance for Needy Families (TANF) saw a 60 percent reduction in immigrant usage, 48 percent in food stamps, and 15 percent in Medicaid (Fix and Passel 2002:15). In addition to limiting immigrants' benefits once they had arrived in the United States, PRWORA aimed to discourage those likely to seek welfare from immigrating in the first place, and shifted much of the financial responsibility for immigrants onto their sponsors (Fix and Passel 2002). Dramatic reductions to the welfare rolls were also found among refugees, with reductions of 53 percent in food stamps, 78 percent in TANF, and 36 percent in Medicaid between 1994 and 1999 (Fix and Passel 2002). In sum, "PRWORA seems to have succeeded in reducing both the number and share of legal immigrants on welfare." Some immigrants, frantic about the loss of their benefits, considered or actually committed suicide (Reese 2005:13). We can certainly understand the implications of the unintended reduction in the welfare rolls as failures of the reforms more broadly. That this legislation enacted "welfare elimination for a group of people currently out of favor and who don't vote" (Reese 2005:186) certainly aided the expansion of measures formerly reserved for the prison system to welfare receipt, including the integration of biometric fingerprinting. Other major reforms to welfare fostered by PRWORA include work requirements for aid recipients, time limits on receiving welfare, mandatory job training, and the use of social programs such as health care and child care as carrots to coerce welfare recipients into forced labor (Gilens 1999; Abramovitz 2000).

Particularly significant about the Clinton-era reforms is the potential they provided for furthering corporate interests. By 1996 the U.S. Chamber of Commerce had made welfare reform one of its highest priorities (Reese 2005). Businesses supported PRWORA in large part because of the tax savings this legislation financed through cuts to social programs. While AFDC itself was not costly, representing only 1 percent of the federal budget, eliminating welfare paved the way for other cuts to social spending in Medicaid and education. This is the business strategy known as "balanced budget conservatism," which aims to reduce government deficits, not through increased taxation of the rich and corporations, but through cuts to state spending on social programs (Reese 2005). Big business also salivated at the possibility of the privatization of welfare. Claiming that they would eliminate governmental inefficiencies and use new technologies and corporate logics to streamline the provision of welfare, big businesses promised to save the state millions. Here we see some of the motivations for corporeal fetishism. Biometric digital maps of the body are incredibly profitable, generating significant value from producing bodies as commodities for the welfare system. When states were granted a free hand in how they could administer reforms many chose to do so by contracting reforms out to private companies (Berkowitz 2001). PRWORA ceded $17 billion in welfare funds to the states, resulting in massive payouts to private corporations (Hartung and Washburn 1998) and making welfare one of the "biggest corporate grab[s] in history" (Berkowitz 2001).

Maximus was the first company to deliver welfare privately, beginning in Los Angeles in 1988 (Berkowitz 2001). Los Angeles was also the first city to adapt biometric technologies to welfare, which it did under the supervision of Maximus. The infamous military corporation Lockheed

Martin was another of the companies responsible for the privatization of welfare. By the late 1990s "Lockheed Martin had landed 15 welfare contracts in 4 states, and Maximus had won contracts for welfare services in almost every state" (Reese 2005:166). The privatization of welfare yielded significant profits for corporations. For example, "Maximus grew from a $50 million operation in 1995 [before PRWORA] to $105 million in 1996, and to $319.5 million in 1999, a 36.8% sales growth over 1998" (Berkowitz 2001:5). In 1999 Maximus was selected by Forbes as one of the ten best small companies in the United States (Berkowitz 2001). With respect to Lockheed Martin, its Information and Services Sector, the branch of the company responsible for welfare, quickly became the company's fastest growing division (Hartung and Washburn 1998).

It is important to note that Maximus, Lockheed Martin, and other companies responsible for the provision of welfare services following privatization were no more successful than state-funded operations. In fact their failures were staggering. Within its first months of operation Maximus had "its operations in the kind of disarray it usually takes government years to achieve" (Berkowitz 2001:7). Maximus had over three thousand complaints in one county alone—almost one complaint for every seven clients. Recipients protested both about disrespectful treatment by Maximus staff and that their cases were not worked in a timely fashion. "With the welfare clock ticking away, not having your case worked in a timely manner becomes critical—bringing recipients that much closer to getting dumped without a safety net" (Berkowitz 2001:6). A legislative audit report in 2000 found $800,000 in questionable spending by Maximus. Lockheed Martin reported similar failures, in one case running over budget by $177 million (Berkowitz 2001). The only way private companies were successful was in dramatically reducing the welfare rolls—although it quickly became clear that deserving recipients were repeatedly dropped from the register—falsely represented by the companies responsible for welfare provision as testament to the success of privatization (Berkowitz 2001). The integration of biometric technologies to welfare delivery provides a useful case study of some of the complications associated with privatization.

It is important to note that cutbacks to welfare were not solely about corporate interests, but were connected to systemic forms of discrimination related to gender, race, class, and disability. Connections continue to be drawn between racialization, gendered identity, (un)willingness to work, and government assistance (Kohler-Hausmann 2007). Thus contempt for welfare recipients stems from racist, sexist, heterosexist, and classist stereotypes about the undeserving poor (Reese 2005; A. M. Smith 2007; Abramovitz 2000). Martin Gilens (1999:6) writes that the news media continually distort welfare, depicting "overly racialized images of poverty" and associating these images with the suggestion that the poor are unwilling to work. Gilens also shows that "Americans who think most welfare recipients (or poor people) are Black express more negative views about people on welfare and are more likely to blame poverty on a lack of effort rather than on circumstances beyond the control of the poor" (206). The imaginary Cadillac-driving "welfare queen" exploited by Reagan in 1976 and utilized by conservative politicians from Clarence Thomas to George W. Bush epitomizes the nexus of discrimination that has been essential to welfare reform, used to justify the rollbacks to federal and state assistance. It is also important to note that cutbacks to welfare rely heavily upon the elevation of heterosexist nuclear family ideals, as marriage is presented as a solution to poverty and queer families are forgotten altogether (A. M. Smith 2007).

Welfare is a gendered program and, with respect to aid for families with children, is in fact the state's sole (and limited) way of recognizing that mothering and homemaking have "social and economic value outside of the patriarchal family" (L. Phillips 1994:1). We saw that welfare reforms particularly targeted newcomers to the U.S., but they were most successful where the caseloads of states were mostly black or Latino (Reese 2005:177). Stereotypes about women of color receiving welfare are essential to cuts to welfare, as are assumptions that women of color should work (179). These types of "commonsense" suppositions were successfully marketed to both working-class and middle-class whites frustrated about "stagnant and declining wages, regressive tax policies, and a rising national deficit" (178) in order to make welfare reform a reality. Thus a marketing strategy of welfare reform based on separating deserving taxpayers from undeserving "tax-eaters" made use of racist, sexist, and heterosexist stereotypes masquerading as home truths.

A Cost-Benefit Analysis of Adding Biometrics to Welfare

Seeing the contracts valued at hundreds of millions of dollars awarded to other companies in the course of welfare privatization, biometric companies like Unisys, Identix, Fingermatrix, and Morpho attempted to jump on the welfare bandwagon. Whether quoted in the media or testifying in congressional debates about the merits of biometric identification for welfare recipients (Associated Press 1996; S. Scott 1996; U.S. Legislature of the State of Texas 1995), biometric companies declared that the primary justification for adding their technologies to social service provision was that these new identification technologies would save the state money by eliminating fraud.[3] Before proceeding, it is necessary to problematize the notion of welfare fraud, given that in many cases welfare is not a living wage and does not meet a recipient's most fundamental human needs for food and housing (Mosher et al. 2004; Mirchandani and Chan 2008; Kohler-Hausmann 2007; Abramovitz 2000; Gilliom 2001). As a result most fraud occurs because welfare recipients work under the table or receive small, forbidden sums outside the system in order to meet their basic subsistence needs. However, even if we take the state's definition of fraud at face value and refer to attempts by recipients to meet their needs for food and housing any way they can as defrauding the system if they go outside the confines of their welfare checks, the only type of fraud that the integration of biometric technologies to welfare delivery aims to address is duplicate-aid fraud. Duplicate-aid fraud occurs when a person signs up more than once to receive benefits, usually using a fake name or identification.

CALIFORNIA

As California was the first state to implement biometric identification programs for welfare recipients and has one of the largest and longest-running systems in the United States, it provides a useful starting point from which to conduct a cost-benefit analysis. California already had intensive and expensive antifraud measures, including a Fraud Bureau that employed thirty people exclusively dedicated to pursuing alleged cases of welfare fraud. The state added to existing measures a biometric fingerprinting program that was applied broadly to all recipients. For example, in addition to biometrically fingerprinting adults applying for aid, if an adult applied for welfare support on behalf of a child, the child's entire household had to provide two digital fingerprints and a photo, whether they were receiving aid or not (Howle and Hendrickson 2003:10). If any adult living in the house refused to do so, the entire household became ineligible for support (8). California was also the first state whose biometric fingerprinting system was audited for effectiveness. The audit raised a number of concerns, including about the cost implications of the system.

The primary motivation for implementing biometric identification programs for welfare recipients in California was the elimination of duplicate-aid fraud. And yet the audit found that California's Department of Social Services implemented biometric additions to welfare without determining how much duplicate-aid fraud was occurring in the state (Howle and Hendrickson 2003; J. Davis n.d.). This is an unusual economic model, in which savings are projected regardless of whether or not losses have been incurred. As Jim Davis, then-director of Computers for Social Responsibility, stated, the general manager of California's Department of Social Services simply asserted that the department "believe[s] that there is a substantial number of persons who are receiving duplicate aid." As Jim Davis paraphrased the remark, "Since [the department] doesn't know, there therefore must be a 'substantial number.'" Thus biometric programs were implemented as a cost-savings device when there were no figures substantiating whether or how much money was being wasted.

Existing measures for determining duplicate-aid fraud in California before biometrics were implemented suggest that the numbers were low (Howle and Hendrickson 2003). Following the integration of biometric technologies in the state welfare system, according to figures received by the Computer Services Division of the Los Angeles County Department of Public Social Services, "After six months of use, all of 11 cases had been closed because duplicate sets of prints were found on the system, although it is not known how many of those cases were considered fraudulent behavior" (J. Davis n.d.). Ultimately the majority of fraud cases found by the biometric system turned out to be the result of administrative error (Howle and Hendrickson 2003). Given the high costs to develop and implement the system—the state of California paid $31 million in initial start-up costs and $11.4 million every year to run its biometric system—coupled with the low amount of fraud occurring in the state, California's biometric program did not begin to break even, let alone save the state money (Howle and Hendrickson 2003). Despite the devastating biometric failures yielded by California's program, this was the model that had already been used to rationalize the expansion of biometric technologies to welfare programs in other states. Given the results in California, it is not surprising that other biometric additions to welfare also failed to save money.

Texas, which explicitly modeled its system after California's AFIRM system (US. Legislature of the State of Texas 1995), also yielded problematic financial results. Again, although fraud was given as the reason for adding biometric technologies to Texas's welfare system, numbers were not furnished to indicate how much money the state was losing as a result of welfare fraud. Still the Department of Social Services remained confident that biometric technologies would save the state money: "This technology will cut out that type of problem. We don't know how much duplicate benefits there are, but we won't know until we try it" (S. Scott 1996).

A study conducted by researchers at the University of Texas at Austin concluded that the Lone Star Image System pilot project was ineffective and expensive. Based on research from October 1995 to May 1997, the report concluded that most welfare fraud could not be attributed to individuals signing up for double benefits, the only type of fraud designed to be caught by the integration of biometric technologies to welfare delivery. Conducting a cost-benefit analysis, the study found that the system cost $1.7 million in its first seven months of operation and yielded no savings whatsoever ("UT Austin Researchers Determine Lone Star Image System Project Did Not Prevent Welfare Fraud" 1997). The biometric fingerprinting program was unlikely to yield any savings even if there was significant fraud in part due to the low level of welfare benefits available in Texas. Texas has the lowest welfare rates in the country; the state doles out a paltry $188 a month for a family with three dependent children, compared to almost $700 in California ("UT Austin Researchers Determine Lone Star Image System Project Did Not Prevent Welfare Fraud" 1997). Even had there turned out to be duplicate-aid fraud, it would have been difficult for the state's biometric program to recoup any significant savings.

Despite the complete failure of the pilot project, the project was expanded. The state awarded Sagem Morpho a contract to biometrically fingerprint welfare recipients from 1997 to 2007 ("Peddling Welfare-Privatization Boondoggles" 2007). Predictably the expansion of the failed pilot project was not a success. The debate was taken up in 2001 by Congressman Glen Maxey, who argued that the program was "pork" and claimed that "the contract had cost the state $16 million over a five-year period that produced nine welfare-fraud prosecutions (or $1.8 million per prosecution)." Describing the integration of biometric technologies with the administration of food stamps, an internal memo in the Texas Health and Human Services Commission said that "the state spent $12 million on fingerprinting services that detected $59,000 in fraud." In response Sagem quadrupled its lobbying expenditures, to a total $1,125,000. Arguing that the small number of welfare fraud cases demonstrated the success of the system rather than its failure, Sagem claimed that fraudsters were clearly being prevented from collecting their benefits, as one could see from the small number of convictions. Sagem highlighted the importance of biometric identification to deterrence. As one of the company's employees said, "It's hard to measure deterrence. . . . It's almost like saying Fort Knox has an incredible security system. Nobody's ever broken in. Why should we be paying for the security system?" ("Peddling Welfare-Privatization Boondoggles" 2007). Ultimately Sagem Morpho was successful in convincing the state to keep its program. The system now costs Texas $2.5 million annually in operation costs (Price 2005).

ILLINOIS

As in California and Texas, prior to its adoption of biometric technologies for welfare, Illinois did not conduct any systematic study of the costs of duplicate-aid fraud. The state's Department of Social Services asserted, "Typically, duplicate cases are found serendipitously rather than systematically. The actual extent of multiple case fraud is not known" (U.S. State of Connecticut Department of Social Services 1999). In order to determine how much fraud was occurring, the state took a figure proposed by the U.S. Congress, which estimated the national cost of all fraudulent welfare schemes at $25 billion annually. Of course this figure represents all types of fraud, not just duplicate-aid fraud. Despite this, the Illinois Department of Social Services simply estimated the state's welfare caseload size to be 1/25th of the nation's welfare caseload and determined that "the Illinois portion of these schemes could be in excess of $1 billion" (Beckwith 1999).

Ultimately, as in Texas and California, the program in Illinois failed to find duplicate-aid fraud. In fact as of 1999 "only one multiple enrollment has been found" (Beckwith 1999). Although the state acknowledged that individual program savings were insignificant, it expected savings to result from deterrence: "One advantage of using biometric identification as a case eligibility requirement is its potential for cost savings through deterrence. This determination of actual savings is elusive and difficult, if not impossible, to measure accurately, particularly against the backdrop of recent federal and state welfare reform policies and the extremely favorable economic conditions in the State of Illinois." Despite the economic pseudoscience justifying the program, Illinois's biometric fingerprinting program remains in place (Accurate Biometrics n.d.), although its retinal scanning program was discontinued (Beckwith 1999).

NEW YORK

New York has one of the largest biometric fingerprinting programs for welfare recipients in the world (Nawrot 1997). It too has run into serious difficulties. Again, the primary reason given for New York's system was fraud deterrence, but again, no data were collected to determine how many cases of duplicate-aid fraud actually existed prior to the state's adoption of biometric technologies. New York City itself spent $40 million to $50 million to implement biometric enforcement, with the hopes of saving a projected $250 million (Firestone 1995:1). Initial savings projections for New York State based on deterrence were enormous. The program was credited with more than 38,223 people being dropped from the welfare rolls in the first two years of the system's operation, leading to claims that the program had saved the state more than $297 million in the first two years (Bernstein 2000).

In 1997, two years after New York began biometrically fingerprinting welfare recipients, the state commissioned a study by research consultants for Maximus and Macro International to audit the system. Although the state spent $658,000 conducting the study, its results were suppressed. In fact New York's Department of Social Services moved quickly to discredit the findings, saying that the study was both outdated and flawed (Bernstein 2000). In 2000, three years after the study was conducted, the *New York Times* managed to obtain a copy. The study found that biometric fingerprinting made no appreciable difference in the dropout or approval ratings of welfare recipients, suggesting that the biometric identification of welfare recipients had no impact on deterrence (Bernstein 2000). As of 1997 only 172 cases had been dropped as a result of fraud, making the expenses per case monumental (Nawrot 1997). The *New York Times* interviewed experts in the field, including Paul Sticha, who reviewed biometric applications to welfare on a national level for the federal government. Sticha upheld the study's results (Bernstein 2000).

Still the state continued to claim huge savings because thousands of people were dropped from the rolls following the implementation of biometric identification. Although only a fraction of those dropped from the rolls were dropped as a result of duplicate-aid fraud, New York's Department of Social Services confused the issue by conflating those dropped from the rolls for any reason with those dropped for fraudulent activity (Nawrot 1997). For example, 712 people were forced off the welfare system because they refused to be biometrically fingerprinted. This does not mean that they were necessarily committing fraud, yet the state interpreted any reason for non-reenrollment as a success (Nawrot 1997). The study's results were never made public, let alone heeded, and New York's program was ultimately expanded (Bernstein 2000).

Other states abandoned biometric identification programs for welfare recipients altogether because of their expense and failure to reduce costs. Maryland's program was ultimately cut (Maryland General Assembly 1998). The state of Maryland found that fewer than 4 percent of cases were due to fraud, meaning that, at best, only sixty people would be cut from the rolls (Howle and Hendrickson 2003:14). North Carolina's program was also abandoned due to the expense (Newcombe 2001). Given the multitude of other aggressive methods that states use to test for fraud, including home, school, and work visits, computer matching to make sure an individual is not enrolled twice, and the collection of enormous amounts of personal data (Howle and Hendrickson 2003:15), it is difficult to justify such an expensive and ineffective system.

Significance

If these systems fail to save the states money by eliminating fraud, what do they accomplish? Biometric companies made significant profits by expanding to welfare. In biometrically identifying welfare recipients, an industry with little direction and no obvious market had found a rich new group of clients in the form of state departments of social services. Although biometric companies failed to eliminate fraud, they were successful at cutting people off the rolls. The welfare roster was not reduced because recipients were afraid of being caught committing fraud, but for a much more complex set of reasons, including immigrant and refugee status, disability, and more generalized anxieties around information sharing.

Measures refined for surveillance in the prison industrial complex routinely are expanded to the welfare system, including anonymous snitch lines for suspected welfare abuse; "one strike" policies for welfare fraud, resulting in permanent ineligibility for benefits; and increased search provisions, allowing welfare caseworkers to inspect recipients' homes without notice (D. E. Roberts 1997, 2001; A. M. Smith 2007). Those receiving welfare have long borne the stigma customarily associated with criminalized acts. As Virginia Eubanks (2006) documents in her study of the use of new information technologies in the surveillance of welfare receipt, welfare compels recipients to reveal every part of their personal lives, similar to the prison industrial complex. One woman Eubanks interviewed notes, "Every part of your life, everything about you, is available. The system—that person that you're talking to—knows everything about you" (94). John Gilliom (2001:28) asserts, "The welfare poor are subject to forms and degrees of scrutiny matched only by the likes of patients, prisoners and soldiers."

Investigators "routinely order [welfare] applicants to empty their pockets, then flip through their wallets and personal possessions, demanding to know the identity of every name they come across" (Firestone 1995:1). Importing measures from the prison system to welfare is also about the expansion of the war on the poor, and in particular the war on poor people of color (Sudbury 2004; A. Y. Davis 1998, 2003, 2005; A. M. Smith 2007; D. E. Roberts 1997), as the same communities are again targeted for criminalization and harassment by the law. Reforms to welfare intensified and expanded procedures that explicitly criminalized recipients.

Adding biometric fingerprinting to welfare is part of the process of the expansion of biometric technologies to new markets. Refined over a period of "20 years for its obvious first customer, law enforcement agencies" (Steinberg 1993:6), biometric fingerprinting development was initially driven by the FBI. The technology used in the welfare system is identical to the finger-imaging technology used to verify the identity of prisoners (Adelson 1994:4). Despite these clear links to criminalization, biometric companies continue to try to shake this association in order to ensure expanded markets for their products. One way is to use the rhetoric of science to claim that biometric finger imaging cannot be confused with ink fingerprinting: "That's manual fingerprinting with ink. This is a very hi-tech system that is a clean process and a dignified process" (McLarin 1995b). Another way that the industry attempts to distinguish the fingerprinting of welfare recipients from criminalizing processes is by asserting that welfare bureaus take only two prints, while prisons take ten. Yet despite these attempts to distinguish biometric fingerprinting from criminalizing processes, the relationship between the finger imaging of welfare recipients and the surveillance and criminalization of welfare is unmistakable, not least because biometric fingerprinting was explicitly avoided in other domains due to the stigma surrounding fingerprinting as a tool of law enforcement. Interestingly the *New York Times* asserted that "most welfare fraud is not done by welfare recipients, but by providers, including doctors and landlords" (Hauppage 1993:1). As a result of the high incidence of provider fraud, lawmakers in Connecticut debated an "amendment that would have required doctors and others who supply goods and services to welfare recipients to be fingerprinted" (Associated Press 1995). Not surprisingly this proposal was rejected, revealing that more powerful citizens are able to reject a criminal classification, while those living at the margins are not.

When biometric programs were introduced to welfare, there were already existing connections between welfare and law enforcement (Little 2001; Reese 2005; A. M. Smith 2007; J. Davis n.d.). Biometrically fingerprinting welfare recipients deepened those connections as information sharing between the two increased. Eubanks (2006:96) documents cases of welfare recipients who cannot get access to public housing as a result of teenage infractions that remain on their electronic records, allowing caseworkers to use criminalized behavior to prevent those living in poverty from attaining the support to which they are legally entitled. In this way, the addition of biometric technologies to welfare is also a narrative about the broadening of existing systems of state surveillance while making money for private corporations (Monahan 2010:120). Moreover Clinton established the Criminal Child Support Task Force, charged with facilitating information sharing between welfare and law enforcement. Thus fathers of children receiving TANF are now tracked down by the police if they fail to make payments, whether they can afford them or not (Justin 1997). Welfare recipients also are obligated to provide their home address to law enforcement as a result of this program (A. M. Smith 2007). Despite earlier claims that the biometric information collected from welfare recipients would remain private, the fingerprints of welfare clients are being made available to other government agencies. In Massachusetts law enforcement officials investigating crimes were allowed to subpoena welfare fingerprint records (Wong and Phillips 1995:1). The state of New York went further in allowing "social services officials to pass on to law enforcement officials cases of fraud revealed through the finger-imaging program, which they had not been allowed to do under the earlier law initiating the program for *Home Relief* recipients" (Fein 1995:26). In those states where biometric information is shared between agencies, the integration of biometric technologies into welfare delivery further increases the likelihood of recipients coming into contact with the prison system.

THE CREATION OF NEW CATEGORIES OF DISABILITY

Perceiving bodies as no more than a series of ones and zeroes fails to attend to the complex webs of power in which bodies are situated and the different needs those bodies may have. Part of the reason that biometric programs resulted in such significant drops in the welfare rolls is because these programs to test welfare clients for fraud have a disproportionate impact on those living with mental health issues. The connection between welfare receipt and poor mental health is well documented (Zabkiewicz and Schmidt 2007; Zaslow et al. 2006; Rosen et al. 2003). National surveys indicate that people with physical and mental disabilities receiving welfare constitute approximately 36 to 44 percent of adult recipients (Reese 2005:14). If one looks solely at mental health issues, these individuals represent approximately 25 percent of all welfare recipients (Zedlewski and Alderson 2001; Polit and Martinez 2001). The rates of welfare recipients suffering from anxiety disorders are almost twice as high as in the general population (Derr, Douglas, and Pavetti 2001).

The large number of people with mental health disabilities receiving welfare is no accident; the original intention of welfare was to provide for those who do not easily fit into an able-bodied capitalist framework that emphasizes productivity and efficiency. As community workers argue, many recipients with mental health issues "have enough trouble just getting out of bed in the morning. Requiring them to follow the complicated steps allowing them to be biometrically fingerprinted could cause them to drop off the rolls" (Monsebraaten 1996). Speaking specifically about welfare recipients living with anxiety disorders, advocates argued that "many of these people are already paranoid. . . . To ask these people to surrender their fingerprints to the welfare bureaucracy could put them over the edge. Many would simply refuse and drop off the system altogether." Nor can recipients' fears about the ramifications of being biometrically fingerprinted be labeled wholly paranoid, given the ways biometric information is being shared between welfare and law enforcement agencies. To date, no provisions have been made for those welfare recipients who are afraid to be fingerprinted; people are simply categorically denied benefits (U.S. Social Security Administration, Office of the Inspector General 2000; J. Davis n.d.). As a result of biometric identification programs, welfare recipients with mental health issues are likely to drop off the rolls ("Frequently Asked Questions: Ending Finger Imaging" 2003; Precious 1992).

IMPLICATIONS FOR IMMIGRANTS AND REFUGEES

We saw that welfare reforms mandated by PRWORA deliberately targeted immigrants and refugees to the United States, increasing the state surveillance of newcomers in a process of "crimmigration" (T. Barry 2010). In addition to PRWORA's explicitly forbidding immigrants from obtaining welfare benefits for their first five years in the country (Reese 2005:3), welfare-to-work requirements and complicated new procedures dealt a special blow to those whose first language was not English. More generally, confusion about the rules governing welfare receipt had a chilling effect on immigrants and refugees who feared that their status would be adversely affected if they were to obtain welfare, even when it was permitted (Fix and Zimmerman 1999). The addition of biometric technologies to welfare also had a significant impact on immigrants and refugees.

A useful place to examine this phenomenon is California, the first state to implement biometric additions to welfare, and a state that has a high proportion of newcomers to the U.S. California saw one of the most dramatic reductions to the welfare rolls when looking to immigrant participation in these benefits (Fix and Zimmermann 1998). Immigrants and refugees were disproportionately affected by biometric additions to welfare provision for many of the same reasons discussed earlier. In addition immigrant communities repeatedly expressed anxieties that welfare agencies would share biometric information with immigration officials (Kruckenberg 2003).[4] Parents' fears about being biometrically fingerprinted resulted in their children going hungry, as some chose to drop out of the food stamp program (Kruckenberg 2003).

A document describing the implications of adding biometric fingerprinting to welfare receipt reported:

> Citizen or legal immigrant children of undocumented immigrants are most vulnerable when it comes to finger imaging. In many cases, even an undocumented parent who cannot receive food stamps is required to give his or her fingerprint in order to get benefits for their child. Undocumented immigrants tend to be especially fearful that the fingerprint will be used to take action against them, when legally it cannot. As a result, finger imaging is a strong deterrent to immigrants who would otherwise enroll their children in the food program. (Kruckenberg 2003)

Requiring biometric fingerprinting of welfare recipients in fact causes immigrant and refugee children to go without food, due to concerns about the way that the information might be shared ("Frequently Asked Questions: Ending Finger Imaging" 2003).

Requiring biometric fingerprinting adds another complicated and bureaucratic step: completing forms in English that many immigrants cannot read or fail to understand ("Frequently Asked Questions: Ending Finger Imaging" 2003; Fix and Zimmermann 1998). Thus biometric fingerprinting increased existing anxieties that receiving welfare would interfere with immigrants' applications for permanent residency, continuing the chilling effect on applicants to welfare (Howle and Hendrickson 2003; "Frequently Asked Questions: Ending Finger Imaging" 2003).[5] Biometric finger imaging programs also interfered with outreach attempts to enroll those who would be eligible for food stamps but did not know about the aid program. Prior to biometric finger imaging, welfare workers would go to nontraditional locations, including "schools and health clinics," to enroll eligible people in the program (Kruckenberg 2003). Since recipients now had to be fingerprinted in addition to many other requirements, welfare workers were expected to take portable fingerprinting machines to their outreach visits. But the machines were heavy to carry and hard to set up, so welfare workers instead decreased their outreach efforts.

Clearly the addition of biometric technologies to welfare only served to intensify existing forms of discrimination, as immigrants were excluded from the benefits of citizenship offered to their U.S.-born counterparts. The targeting and exclusion of immigrants from welfare is in keeping with the ways that people of color are disproportionately targeted by welfare cutbacks (Mirchandani and Chan 2008; A. M. Smith 2007; Eubanks 2006).

Conclusion

In *Backlash against Welfare Mothers* Ellen Reese (2005:19) asks, "If restrictive welfare policies cause such misery, why is political support for them so strong?" Although biometric additions to welfare clearly fail to provide recipients with access to the benefits to which they are entitled, they succeed even when they fail. As part of the larger project of privatizing welfare, biometric technologies made millions of dollars for an industry searching for profits and a sense of purpose. In addition they fit into a political climate in which politicians hoped to gain currency by criminalizing immigrants to the U.S. and poor people, in particular poor women, of color. Integrating biometric technologies in welfare service provision succeeded in forcing dramatic drops to the rolls, although not as a result of welfare fraud. Instead millions of dollars were squandered on a technological solution to the imaginary problem of duplicate-aid fraud.

Perhaps the most egregious failure of the integration of biometric technologies into welfare service delivery is the waste of scarce state resources at a time when there were increasingly fewer dollars to address poverty in the U.S. Money was thrown away on a technological solution to a complex social problem at the same time as a combination of neoliberal policies and racist and sexist backlash had whittled away state resources to address poverty. Biometric technologies were integrated into welfare provision at the expense of depriving poor people of access to the most basic means of subsistence: food and housing. Can adding biometrics to welfare be deemed a success given the consequences for the intensification of the surveillance of recipients, including the disproportionate impact of biometrics on immigrants, refugees, and people with disabilities? The answer is a resounding no.

four BIOMETRICS AT THE BORDER

Biometric technologies are central to the remaking of borders. The productive impact of biometric technologies on borders and border control is especially revealing at the U.S.-Canada border, given its dramatic transformation after 9/11. That's when representations of the boundary between the U.S. and Canada changed from an unguarded, symbolic line between friends to a dangerous fissure in U.S. national security policy. It is a space to study the ways that these new identification technologies are mobilized in the construction of the post-9/11 state.

Although there has been a proliferation of cultural studies and feminist analyses of the southwestern border of the U.S. (Lugo 2000; Anzaldúa 1987; Ono and Sloop 2002), the border with Canada has received less scholarly attention, with notable exceptions (Salter 2007; Drache 2004). Imagined to be a space less in need of securitization and surveillance before 9/11, after 9/11 this border is undergoing a shift, a transformation to which biometric technologies are deemed essential. As a result this border represents an excellent case study to examine how biometric discourse is involved in reshaping territories, borders, bodies, and identities. Historically Americans saw Canadians as white, middle-class, nonthreatening visitors from the North. However, as all categories of racial identification are inherently unstable, this characterization proved to be no exception. Since September 11 a shift in discourse is transforming this boundary from an unguarded space to a reified line whose openness is now described in the media as a "luxury" (Nuñez-Neto 2005). As the United States ceases to describe its neighbor as the Great White North, Canadian bodies are reenvisioned as potential pollutants, threatening to contaminate the U.S. nation-state. Two incidents are central to this transformation, both of which spurred calls for new technologies able to "outsource" the border away from U.S. territory. The first involves the initial rumors that the 9/11 terrorists entered the United States through Canada. This original understanding has profound ramifications for reconfiguring border agreements between the two countries. The second incident involves the arrests of the "Toronto 18" and the consequences of these arrests for the management of the border.

Both incidents are used to justify post-9/11 border policies calling for biometrics able to make newly suspect Canadian bodies visible at the border. Studying the adoption of biometrics at the border provides "before" and "after" snapshots of these new identification technologies as they are enlisted in the redefinition of the spatial and cultural landscape of the country's northern boundary. Although the border between Canada and the United States is described as self-evident, in fact attempts to use biometric technologies to make the border visible demonstrate that the line between the two nations is not as straightforward as it is imagined to be. Like the construction of biometric technologies themselves, the process of making the border visible depends upon practices of inscription, reading, and interpretation that are assumed to be transparent and yet remain complex, ambiguous, and inherently problematic.

Locating the Border

Stretching more than five thousand miles, the border between the United States and Canada is the longest in the world. At close to 680 billion Canadian dollars, the value of bilateral trade between the two countries is the world's highest. Using pictures of helping hands or images of an unguarded, friendly place, media and state depictions of the border describe it as the longest unmilitarized boundary on Earth.

Canada occupies an exceptional place in the U.S. national imaginary. This is partly the result of the significant trade relationships. The early 1980s marked a shift in the Canadian business community's relationship to cross-border markets and its growing assertion that protectionist policies were denying it access to lucrative international markets. Canada's business community organized and began to lobby in support of liberalized trade agreements that would allow it greater access to the American market (Gabriel and MacDonald 2004). In 1984 the Conservative government of Brian Mulroney came to power and, working with the Canadian business community, negotiated the Free Trade Agreement, which the two countries signed in 1988 (Brunet-Jailly 2004:7). The North American Free Trade Agreement (NAFTA), which eliminated tariffs on trade between Canada, the United States, and Mexico for ten years, was signed in 2002. NAFTA significantly increased Canadian economic dependence on the U.S. From 1994 to 2002 trade doubled between the two countries (Fry and Bybee 2002:6). By 2004 over 85.1 percent of Canadian exports were destined for the U.S. (Brunet-Jailly 2004:7).

Canada's greater dependence on trade with its southern neighbor leads to an asymmetrical power relationship. However, trade with Canada is not inconsequential to the United States, as Canada is its largest trading partner (U.S House of Representatives 2006). Canadian imports from the U.S. nearly doubled in size between 1993 and 2000 (Hufbauer and Vega-Canovas 2003). Canada is the leading export destination for thirty-nine of the fifty states (Government of Canada 2007). Ontario is Michigan's largest trading partner overall (Brunet-Jailly 2004). Moreover American interest in Canada is not limited to trade. Canada has the second largest known oil reserves in the world (Government of Canada 2007) and thus is a significant source of oil as well as gas, hydroelectricity, and uranium (Gibson 2007; Government of Canada 2007). It is also likely that the U.S. will depend increasingly on Canadian water in the future, as its own supply is insufficient for its growing needs and Canada has the largest amount of fresh water in the world (Shrybman 2007).

This significant trade relationship contributed to keeping the border open and relatively unguarded prior to the dramatic changes wrought by the attacks on the World Trade Towers and the Pentagon. However, the narrative of the "longest undefended border in the world" also relied upon the imagined whiteness of the Canadian state. This cultural location is as important as trade to understanding the nature of Canada's special friendship with the United States.

Canadian Exceptionalism

The privilege that Canada historically enjoys with respect to the United States is connected to the racialization of the Canadian state; that is, Canada's "special friendship" with the United States is determined in part by its imaginary whiteness. But since 9/11 this narrative has undergone a period of transformation.

Evidence of Canadian exceptionalism is made clear through a comparison of the differences between America's treatment of Canada and its treatment of Mexico. In her dissertation on disease and the U.S.-Mexico border, the communication theorist Maria Ruiz (2005:6) found that the metaphors used to describe the southwestern border consistently represented it as a "frontline dividing 'us' from 'them.'" Mexico itself is cast as a source of illegal pollutants threatening to infect the U.S. national body. In contrast, the media represent Canada using metaphors like the "Great White North" and depict the U.S.-Canada border as an "undefended" line between "friendly neighbors" (Gorham 2006). As discourse has real-life consequences for material boundaries (Ono and Sloop 2002), these discursive differences in the representation and racialization and surveillance of the two countries result in two very different sets of rules for Canadians and Mexicans with respect to the United States.

One manifestation of this contrast is the differential treatment and scrutiny of those Mexicans and Canadians who enter the U.S., as they are entitled to do, without obtaining a visa. Canadian nationals may stay in the U.S. without a visa for up to six months (U.S. Department of State 2005–2006), compared to seventy-two hours for Mexican nationals. During that time Mexican nationals cannot move beyond a proscribed "border zone" which extends only twenty-five miles from the U.S.-Mexico border into California, Texas, and New Mexico and seventy-five miles into Arizona (U.S. Consulate General in Ciudad Juarez 2007). There also are significant differences in the money, surveillance, and security personnel devoted to the northern and southern borders (Andreas and Biersteker 2003). By the end of the 1990s the border with Mexico was heavily implicated in intensified regimes of surveillance. There were more border agents in Brownsville, Texas, than along the entire five thousand miles of the U.S.-Canada border (Andreas 2005). Even after 9/11 there are only 1,013 agents deployed along the northern border (U.S. House of Representatives, Minority Staff of the Committee on Homeland Security 2005) in comparison to the 12,000 officials stationed along the southwestern boundary (Meyers 2006). Nuñez-Neto (2005) cites a report by the Congressional Research Service published in 2005:

> There are significant geographic, political, and immigration-related differences between the Northern border with Canada and the Southwest border with Mexico. Accordingly, the [U.S. Border Patrol] deploys

a different mix of personnel and resources along the two borders. Due to the fact that over 97% of unauthorized migrant apprehensions occur along the Southwest border, the [U.S. Border Patrol] deploys over 90% of its agents there to deter illegal immigration.

Although the U.S. increased security along its northern border following 9/11, these investments are a fraction of those devoted to surveillance at the Mexico border. The Bush administration planned the addition of hundreds of miles of fence as part of a $1.2 billion deal to improve border security (Reuters 2006a). The Obama administration also supports border security, although its version relies on different technologies (Hsu 2009). Canadian exceptionalism is clear when compared to the militarization of the Mexican border. In the case of Mexico, an extensive trade relationship is insufficient to guarantee the same privileges that Canadians enjoy.

Canada's privileged place results in part from American representations of Canada as a safe refuge, a pristine wilderness, or a backward haven of beer, hockey, and maple syrup. For example, in the science fiction television series *Dark Angel* (2000), the United States has suffered a massive technological failure referred to only as "the pulse," a giant electrical surge that destroyed the nation's infrastructure. The show reliably ends when those who are rescued are safely transported to Canada. In *Dark Angel* Canada is cast as a mystical retreat, and the border is represented as the gateway to a better elsewhere.

In a related vein, Weird Al Yankovic's song "Canadian Idiot" captures the dual stereotypes of Canada as safe haven and hilarious hotspot on the cutting edge of the Stone Age. Relying on the familiar, though contradictory, rendering of the beer-drinking, hockey-mad, overly polite, frostbitten Canadian Mountie, the song simultaneously mocks Canadian earnestness ("They treat curling just like it's a real sport") and the Canadian climate ("Snow's what they export"), while paying homage to the continued Canadian commitment to the welfare state in the form of socialized medicine, a strategy adopted by Michael Moore in *Sicko* (2007).

These images of the Great White North all depend on the imagined whiteness of the Canadian state. Michael Moore's film *Canadian Bacon* (1995) best captures Canada's racialization in the U.S. national imaginary.[1] *Canadian Bacon* begins as the cold war draws to a close, an end that produced an unprofitable peace for the newly elected president. As anger at the closing of military plants leads to plummeting ratings in the polls, the president turns to his aides in desperation. They suggest a war, guaranteed to unify the country and boost his ratings. The only question that remains is who to fight. The aides offer a number of suggestions, but the pickings are poor; from aliens to dead revolutionaries, each possibility seems more unlikely than the last. In a prescient piece of filmmaking, a war on international terrorism is suggested and then summarily rejected by the president as ludicrous. Randomly flipping through channels, one of the president's aides comes across footage of a riot started at a hockey game by an American who insulted the quality of Canadian beer. As hypermasculine hockey players pummel American fans, the president's aide has an idea: Why not start a war with Canada? The film's depiction of the labor required to turn Canada into an enemy reveals a good deal about American ruminations on the Great White North. Describing the freshly minted "Canadian threat," the foreign affairs officer in charge of Canada says, "We used to be blind to Canadians. We thought—they're just Canadians—practically the fifty-first state. We admired them—clean streets, no crime, no minorities." When the president's aide agrees, asking, "Yeah, how'd they do that?," the response is "No slavery," earning a reply of "God, they're smart." When the president hesitates to enter into battle with Canada, asserting that the American people will never buy this war, his staff is initially in agreement, offering as an added barrier, "Hell, they're whiter than we are!" Yet the aides persist, telling the president to "wake up and smell the snow": "The American people, Mr. President, will buy whatever we tell them to. You know that." Shortly after, preparations for war begin. An American TV station highlights the sneaky ability of Canadians to blend into the United States, which one of the president's aides summarizes as "Fact: Canadians cross our boundaries and walk among us—undetected." Here the film highlights the imagined whiteness of the neighbors to the North, representing Canadian bodies as "undetectable" as a result of their unmarked racialized identities. Eventually a corporate media coup transforms Canada into the reviled other, and Americans eagerly jump on the bandwagon to level Toronto. The end of the film reveals the bogus nature of Canada-as-military-threat, and the bombing of Toronto's CN Tower is narrowly avoided. The reliance of *Canadian Bacon* on stereotypical representations of Canada as a safe, polite society is a narrative to which the imagined whiteness of Canadians is key.

These depictions of Canada as a more benign, liberal version of the

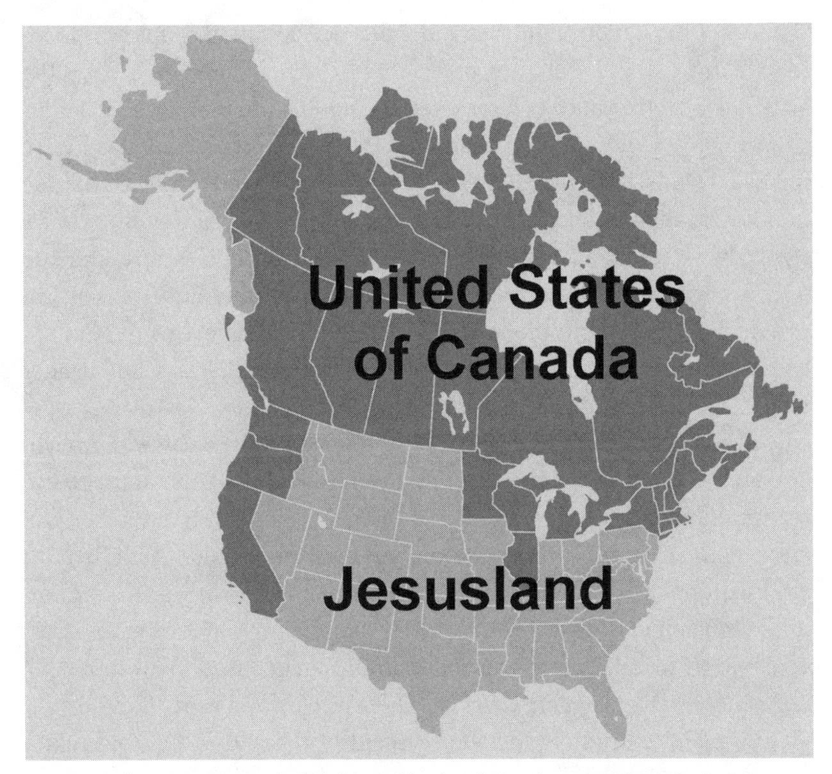

FIGURE 7 This image was included along with other fictional content about the imaginary Republic of Jesusland, created after the U.S. presidential election in 2004. It was found on http://www.uncyclopedia.org, a spoof of Wikipedia.

United States are of long standing. The "red state/blue state" division between the Republican and Democratic parties, first popularized by American media during the controversial presidential election in 2000, was conventional shorthand by 2004. Canada was seen as kinder and gentler than the U.S., without its vicious partisan politics. After Bush's second victory, a map that showed the "blue states" merged with Canada to form "the United States of Canada" circulated widely on the Internet (fig. 7). Variations on this theme showed an "escape route" through the blue states into Canada.

Canada's protective veneer of white privilege in the U.S. national imaginary is no accident. The Canadian state invested considerable labor in inventing itself as a white settler society, relying on the colonization of Aboriginal peoples and on racist immigration policies. Canadian confederation constructed both white French and English settlers as the rightful "fathers of the nation," ignoring the rights of Aboriginal peoples. Policies mandating that Aboriginal peoples could claim entitlements from the state only when resident on reserves helped to contain racialized bodies within the space of the Canadian state itself (Thobani 2000:290). Residential schools and laws that deprived Aboriginal women who married non-Native men of their Aboriginal status were part of a deliberate attempt by the Canadian state to force Aboriginal peoples to assimilate (Obomsawin 2006; Monture-Angus 1999). Imagining its origins as European also had implications for Canadian immigration policy. Certain bodies were not admitted, a process dependent on race, class, and gender identities (Thobani 2000; Razack 2002). In 1921 race was made a specific category upon which admittance to Canada was premised (Thobani 2000:287). The word *race* was removed from the legislation and a clause that prohibited discrimination based on "race, national or ethnic origin, color, religion or sex" was added only in the Immigration Act of 1976–77. Inequities based on race and gender remain. Recently Canadian immigration policy forbade domestic workers from the Caribbean, overwhelmingly female, to sponsor their families in an effort to prevent their permanent settlement in Canada (Thobani 2000, 2007). Such immigration policies promote the containment of othered bodies and maintain the imaginary whiteness of the Canadian state. Sunera Thobani (2000:281) refers to this exercise as "bordering," a violent process in which racialized men and women are excluded from the privileges of Canadian citizenship at the same time as their labor is "put into the service of the national economy." In this way racialized immigrants, particularly women of color, are consistently forced within the boundaries of what the theorist Himani Bannerji (2000) terms "the dark side of the nation."

Canadian "Terrorists" and 9/11

The shift in the place of Canada in the U.S. national imaginary and the increasing inclusion of Canadians in the growing surveillance apparatus began immediately following the attacks of September 11, when several major American newspapers claimed that a number of the 9/11 hijackers had entered the United States from Canada. A story first published by the *Boston Globe* on September 13 suggested that American investigators were "seeking evidence . . . that the hijackers responsible for Tuesday's attacks had slipped into the United States from Canada" (Nickerson and Barry

2001). On September 14 the *Washington Post* asserted that two of the suspected terrorists had definitely entered the U.S. across its northern boundary: "Two suspects in Tuesday's terrorist attacks in the United States crossed the border from Canada with no known difficulty at a small, border entry in Coburn Gore, Maine, which is usually staffed by only one border inspection officer, a U.S. official said today" (Brown and Connolly 2001). The continuing persistence of this rumor and other mistaken beliefs led the Canadian government to make efforts to dispel known falsehoods about Canada's connection to terrorism. In May 2004 the government of Canada launched the website CanadianAlly.com. Aimed primarily at Americans, the website provided a number of facts about the relationship between the United States and Canada and included a "Debunking the Myths" section that roundly denounced the continuing rumor that terrorists had entered from the North:

> True or False: Some of the 9–11 Hi-jackers entered through Canada:
> FALSE
> This is simply not true. In fact, they had all been legally admitted to the United States, as has been confirmed by senior American officials. (Government of Canada 2007)

Yet despite numerous attempts by the Canadian government to extinguish this myth, fears that Canada is a "terrorist haven" continue to spread. Although quickly disproved (Sallot 2001), the rumor of terrorist entry from Canada remains in circulation. American politicians from Secretary of Homeland Security Janet Napolitano to former Republican presidential candidate John McCain continue to repeat this story (Alberts 2009). In 2005, demonstrating that connections were being drawn between Canada and Mexico as related sources of terrorist threats, suggesting the impending Mexicanization of Canada, Newt Gingrich asserted, "Far more of the 9/11 terrorists came across from Canada than from Mexico." Gingrich later retracted this doubly wrong assertion (none of the terrorists came from either Canada or Mexico) when confronted with its inaccuracy by the Canadian ambassador to the U.S. ("Newt Gingrich Sorry for Comments about Canada" 2005). His comments do, however, point to the ways that Canada and Mexico are now being collapsed together as the sources of pollutants to the United States, sentiments that were echoed in a map published on CNN.com (fig. 8).

Although this image and the accompanying story referred to radio-

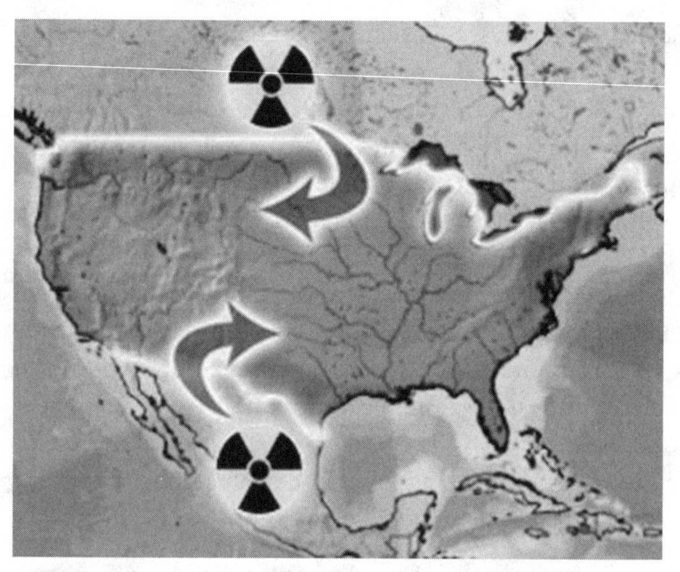

FIGURE 8 This image accompanied an article about two teams of government investigators, who, using fake documents, were allegedly able to enter the U.S. with "enough radioactive sources to make two dirty bombs" (de Sola 2006).

active material that was successfully smuggled across the borders by two undercover agents testing U.S. border security, the image suggests, as Gingrich did, that both Canada and Mexico are now sources of dangerous pollutants to the U.S. American officials expressed outrage at the ease with which illegal materials could enter the country from across either border. Additionally they voiced anxiety about the ease with which Canadians could slip into the U.S. unnoticed, leading to calls for a new technology able to accomplish the task of definitively identifying slippery Canadian bodies at the border.

Why has this mistaken assertion that the 9/11 hijackers came from Canada, disproved almost immediately, nonetheless persisted? Canada-as-terrorist-haven is a useful tool in the justification for attacks on Canada's immigration and refugee policies, and this assault is connected to the post-9/11 racialization of Canada in American discourse. After September 11 U.S. congressional representatives repeatedly "blasted Canada as an unwitting haven for a large number of terrorists, blaming soft immigration laws" (Gorham 2006). Shortly after the attacks on 9/11 members of the U.S. Foreign Service claimed that "Canada's political asylum laws have helped make the country a 'safe haven' for foreign extremists" (Brown and Connolly 2001). Making clear that the perils of Canadian immigration are connected to the racialization of those crossing the border, a commentator in the *Los Angeles Times* argued, "Security controls are famously lax in Canada because politically correct Canadians do not differentiate between 76 year old Madame Dupont coming to visit her grandchildren and bearded young men from Islamic countries" (cited in Andreas and Biersteker 2003:454). Voicing the same problematic views about newcomers to Canada, Douglas MacKinnon (2005), former press secretary to Bob Dole, argued that "the Canadian government not only willingly allows Islamic terrorists into their country, but does nothing to stop them from entering our nation." Analyzing the durability of the myth of the 9/11 hijackers makes clear that the imagined whiteness of the Canadian state is changing. However, the instant that Canadian bodies are no longer racialized as white, we see that the bodies of people of color are falsely collapsed into the category of "terrorist." The transformation and increased surveillance of Canadians in the U.S. national imaginary was accelerated as a result of media and government coverage of the arrests of the Toronto 18.

THE TORONTO 18

In what became known as the "most high-profile anti-terror sweep" in Canadian history (Akkad 2007), the arrest of seventeen terrorist suspects in Toronto in June 2007 furthered the makeover of Canada from friendly neighbor into terrorist haven. Quickly dubbed the "Toronto 17" (a title that caused the media some difficulty when an eighteenth suspect was arrested one month later), the eighteen suspects were ultimately charged by the Royal Canadian Mounted Police with the intent to carry out terrorist attacks in Canada. Although the RCMP refused to name any suspected targets, the media variously reported that the alleged terrorists planned to storm the Canadian Parliament, occupy the building housing the Canadian Broadcasting Corporation, and behead Prime Minister Stephen Harper ("Canada Charges 17 Terror Suspects" 2006). As the media leaked details of the alleged plans, procedural irregularities in the investigation suggested that the planned "terrorist attacks" might be no more than brash claims made by individuals with neither the intent nor the ability to follow through. Errors in the collection of evidence called into question the claim that the group had acquired three tons of ammonium nitrate (a commonly used fertilizer). *Maclean's*, a weekly Canadian current affairs publication, reported that the RCMP relied on two informants in their investigation, one of whom was paid more than $4 million (Friscolanti 2007). This informant, whose name was never published and who has been issued with a new identity, helped the Toronto suspects to purchase the ammonium nitrate from plainclothes police officers. The exchange occurred only after the RCMP substituted a harmless white powder for the fertilizer. Within minutes of the exchange, the RCMP surrounded the group and made arrests. *Maclean's* questioned whether the RCMP and the Canadian Security Intelligence Service had investigated suspicious behavior or had crossed the line into actively encouraging criminal activity in paying their informant to entice the Toronto group to buy bomb-making materials.[2] The second informant on whom the RCMP relied is Mubin Shaikh, a man of relatively dubious character. Shaikh has numerous assault charges to his name and recently acknowledged that he again became addicted to cocaine after the arrests of the Toronto 18 (Friscolanti 2007). Many argued that relying upon Shaikh as the Crown's star witness was problematic given his record, especially as his testimony is so crucial to the case against the Toronto 18 (Austen 2008).

The Toronto 18 showed an unusual and profound lack of knowledge necessary for a terrorist group alleged by the RCMP to have "posed a real and serious threat" with "the capacity and intent to carry out [terrorist] attacks" ("In Depth: Toronto Bomb Plot" 2006). The group failed to notice that their fertilizer had been switched for three tons of harmless placebo. Nor did the alleged terrorists make much effort to blend into their surroundings to avoid detection. Their alleged "training camp," rather than being located in a remote part of Canada where they would garner little attention, instead appears to have been located close to a township in southern Ontario. Eighteen racialized men in camouflage shooting at targets attracted so much attention even before arrests were made that one journalist commented that if this group really was a terrorist organization, they certainly were "second-rate" (Walkom 2006). Charges were dropped against one of the alleged members of the Toronto 18, and charges against two other suspects (both of them under eighteen) were reduced (Akkad 2007). Charges were stayed against seven of the original members, one of whom was found not guilty; ten remain to be tried. The *Globe and Mail* suggested that the reduction (and elimination) of some of the charges suggests that the "accusations levied against at least some of those charged in connection with Canada's most high-profile anti-terror sweep are not nearly as strong as they first appeared" (Akkad 2007). One member of the group, a seventeen-year-old male, was convicted in the first major decision under Canada's tough post-9/11 antiterrorism laws in September 2008 (Freeze 2008). This decision is troubling, especially because even Mubin Shaikh stated that "the youth was not a terrorist and was not privy to the details of any murderous plan" (Leong 2008). Also problematic is that this youth was convicted solely based on his membership in the group, even though there is "no evidence that he planned, or even knew about, any specific plot" (Freeze 2008). Justice John Sproat, the presiding judge, refused to entertain the suggestion that the boy was simply under the sway of a charismatic, self-aggrandizing leader. Nor would Sproat make any allowances for the fact that the group's plans, such as they were, were impossible to attain. That this youth could be convicted and punished harshly despite the fact that it is unclear whether he knew about the crimes that the group may or may not have been planning illustrates the racist implications of Canada's new antiterrorism laws (Freeze 2008). The new laws may be challenged in the courts, as it is unclear whether or not it is constitutional to sentence someone as a terrorist even if he was unaware of the terrorist plot (Freeze 2008; Perkel 2008).

Media reports concerning the Toronto suspects demonstrate dubious attention to due process (Fisk 2006). The suspects are rarely referred to as innocent until proven guilty; instead presumptions of their guilt are regularly reported alongside brief caveats about the importance of assuming their innocence. In her column in the *Globe and Mail* Margaret Wente (2006) writes, "Before I go on, one disclaimer. Nothing has been proven, and nobody should rush to judgment. Meantime, we shouldn't be terribly surprised. The exposure of our very own homegrown terrorists, if that's what these men aspired to be, was both predictably shocking and shockingly predictable." Another *Globe and Mail* columnist, Christie Blatchford (2006), expressed similar sentiments, saying sarcastically, "Why, it's not those young men—with their three tonnes of ammonium nitrate and all the little doohickeys of the bomb-making trade—who posed the threat. No sir: They, thank you so much, are innocent until proved otherwise and probably innocent and, if convicted, it's because of the justice system." In addition to the "ammonium nitrate," the evidence collected included "five pairs of boots in camouflage drab, six flashlights, one set of walkie-talkies, one voltmeter, one knife, eight D-cell batteries, a cell phone, a circuit board, a computer hard drive, one barbecue grill, one set of tongs suitable for turning hot dogs, a wooden door with 21 marks on it and a 9-mm handgun" (Walkom 2006). Although this collection suggested, at best, unclear evidence of the group's plans, photos of the supposed bomb-making "doohickeys" were included in newspaper reports as if the danger in the collected flashlights and barbecue grill were self-evident.

The treatment of the Toronto 18 once incarcerated and awaiting trial raises questions of whether their human rights are being violated. Despite the irregularities in their arrests, many members of the group are kept in isolation, and some are held in solitary confinement for twenty-three and a half hours a day. The apparent motivation behind the isolation is to prevent them from communicating with one another, a security order that their lawyers deemed both cruel and ludicrous, particularly as the accused continue to eat together. One of their lawyers describes their prolonged isolation as a form of torture, and it is currently being contested in the Canadian courts (Leong and Konynenbelt 2007).

In keeping with the media references to the Toronto 18 as "brown-skinned threats" (Appleby and Gandhi 2006) and terrorists "bearded in the Taliban fashion" with "first names like Mohamed, middle names like Mohamed and last names like Mohamed" (Blatchford 2006), media reports following the arrests bear witness to an increase in debates about the dangers of multiculturalism and, predictably, a surge in questions as to the perils posed by Canada's immigration and refugee policies. These discussions are posited as the natural outcome of the arrests, as if the alleged plans of the accused reveal an urgent need to redesign Canada's immigration and refugee system. Calling Canada's immigration and refugee laws into question occurs despite the fact that most of the group members were born in Canada or are longtime Canadian citizens, resulting in a linguistic shorthand for differentiating between these alleged terrorists and the rest of Canadians. The celebrated British foreign correspondent Robert Fisk points out that these media descriptions of the Toronto 18 as "Canadian-born" (S. Bell 2007; Shephard 2007; G. Dyer 2006) rather than plain old "Canadians" is problematic. Both Canadian and American media also describe the Toronto 18 as part of a rise in "homegrown" threats (Kassamali and Ahmad 2006), whom Prime Minister Stephen Harper describes as "[people who reject] who we are and how we live, our society, our diversity and our values" ("In Depth: Toronto

Bomb Plot" 2006). Harper refuses to endorse the suggestion that a terrorist attack in Canada might be in response to Canadian military involvement in Afghanistan, although the arrests were used by the Harper government to justify Canada's continued participation in that conflict. The emphasis on homegrown, "brown-skinned" terrorists represents another step in the transformation of Canadians from benign (white-skinned) neighbors to potential (brown-skinned) threats.

The arrests of the Toronto 18 did not go unnoticed in the United States, where they quickly became a subject of congressional hearings of the Subcommittee on Immigration, Border Security and Claims on the threat posed to the U.S. by Canada. Describing Canada's border as "even more poorly guarded than the southwestern border" (U.S. House of Representatives 2006), the U.S. government points to the Toronto 18 (somewhat paradoxically) as the latest evidence that Canada is a terrorist haven, proving the need to securitize the U.S.-Canada border. Although the arrests of the Toronto 18 might be interpreted as evidence that Canada was getting tough on terror, Capitol Hill and the American media rebuked Canadians for failing to take the problem of terrorism seriously enough. Hearings on whether to implement the Western Hemisphere Travel Initiative, requiring that both Canadians and Americans present passports at the border, consistently referred to the arrests in Toronto as justification for the need to secure the border. In his introduction to the hearings, Republican Congressman John Hostettler hoped that "the arrests in Canada of 17 Jihadists, mostly home grown, will cause the opponents of secure borders to reconsider because the threat won't go away soon" (U.S. House of Representatives 2006). Hostettler further asserted, "We in the United States have a much more clear focus on the problem of terrorism and have moved beyond denial that our own citizens are capable of terrorism. That doesn't seem to be the case north of the border. The brother of one of the men arrested was . . . quoted this week in the Canadian press newspapers as saying 'he is not a terrorist, come on, he is a Canadian citizen.' " With this statement Hostettler pointed to the importance of understanding Canadian citizenship as a newly suspect category.

Witnesses at the congressional hearing described Canada as "heavily infiltrated by terrorists." This claim was directly attributed to "Canada's immigration and refuge [*sic*] system [which] has been a big part of the problem in per capita terms, [as] Canada takes in double the number of immigrants, and three or four times the number of refuges [*sic*], as the

United States." In coupling immigrants to terrorists in this way, it became possible to interpret Canada's larger number of immigrants and refugees as necessarily representing a larger number of terrorist threats.

Some witnesses went so far as to suggest that the Toronto 18 are only the tip of Canada's terrorist iceberg. Janice Kephart, an American security consultant working to match private sector security companies with government clients, refers to the Toronto 18 as only a small part of Canada's "significant terrorist community. About 50 terror organizations [are] actively operating there and about 350 individuals [are] being actively watched. And according to Jack Hoppe, [Canadian Security Intelligence Service] Deputy Director today, Canada's problem is growing. Only 17 were caught in Canada last weekend and we don't even know if they were some of the 350 Canadian intelligence already knew about."

The Canadian press carried a number of angry responses to the subcommittee's characterization of Canada as a terrorist breeding ground. In particular the Canadian media mocked the subcommittee's ignorance of Canada—not least because Congressman Hostettler referred specifically to the enclave of "South Toronto" as "hosts to the radical Imams who influenced the 9/11 terrorists and the shoe bomber." When asked by reporters to describe this neighborhood, Hostettler asserted that South Toronto is "the type of enclave that allows for this radical type of discussion to go on" (Freeman 2006). As Lake Ontario lies south of Toronto, there is no such neighborhood.

Jon Stewart's *The Daily Show* ran a spoof of the arrests of the Toronto 18 in a broadcast segment called "Maple Leaf Rage" (2006):

> I'm very excited to hear that Canada has broken up a terrorist plot. Seventeen terrorists that were planning to blow up targets in Canada. And I just wanted to say to Canada tonight: Congratulations on becoming a terrorist target.
>
> You know, a lot of us didn't think you had it in you. We thought, oh Canada, these terrorists are just going to come through Canada to get to us. . . .
>
> By the way, before any of us get too nervous about the extent of the terror plot, remember with the exchange rate, 17 Canadian terrorists is only about 15.2 American terrorists.

Offering a familiar, dehistoricized view of the "kindler gentler sidecar of a country we call Canada" founded on principles of democracy and equality,

Stewart pretends bemusement at the arrests in Canada. Congratulating Canada on becoming a terrorist threat, he appears both confused and amused by the fact that Canadian involvement in the War on Terror might have made it a terrorist target. Asserting that the terrorist plot was in response to Canada's military intervention in Afghanistan, Stewart makes comic references to the rise in global instability and warfare since 9/11 by telling his viewers that the war against the Taliban "is so two jihads ago." In response to allegations that the leader of the Toronto 18 "reportedly incited followers at a local mosque with sermons filled with hate against Canada," Stewart pretends amazement. Telling his audience that revealing hatred for Canada is innately funny ("You hate Canada? That's like saying 'I hate toast.' It's not the kind of thing that inspires passion in either direction"), he captures the need for Americans to *learn* to see Canada as a potential source of danger to the United States. His pretended difficulties in reimagining the Canadian state echo Moore's film *Canadian Bacon*, reminding us of the work required to accomplish this task. Clearly new tools were going to be necessary to the challenging task of transforming Canadians in the U.S. national imaginary.

Despite the difficulties inherent in repackaging Canada as a terrorist hotspot, both the endurance of the myth that the 9/11 hijackers came from Canada and the rush to blame Canada's immigration and refugee policy for the Toronto 18 demonstrate that Canadian exceptionalism is undergoing a post-9/11 shift, a shift to which new technologies able to make newly threatening Canadians visible at the border as well as able to remove ("outsource") the border from the territorial edges of the U.S. became essential. Perhaps nothing is clearer proof of the transformation of the longest undefended border in the world than the decision of the racist U.S. Minuteman Project to deploy troops along the Canadian border for the first time: "We shouldn't have to be doing this," Simcox, a Minuteman project organizer, told reporters in Washington. "But at this point, we will continue to grow this operation—also to the northern border" (Jordan 2005). Canadian bodies no longer disappear into the unmarked category of benign whiteness, but glow with newly radioactive racialization.

How to actualize the paradoxical project of making newly suspect Canadian bodies perceptible at the border without unduly burdening trade? Despite ongoing attempts in the Canadian media and by the Canadian government to "border" these "brown-skinned threats" from the rest of good old hockey-loving, beer-drinking, white Canadian settler society, it is clear that American ideas about Canada are changing. As Canada ceases to be America's weaker but wholly benign twin, new technologies are needed to revisualize Canadian bodies.

Enter Biometrics

Yes, it's not glamorous. It's not high-tech. It's chain saws and weed whackers. But if you don't get that basic job done, all I know is cameras won't work.

DENNIS SCHORNACK, U.S. COMMISSIONER OF THE INTERNATIONAL BOUNDARY COMMISSION, THE INTERGOVERNMENTAL AGENCY RESPONSIBLE FOR MAINTAINING THE U.S.-CANADA BORDER

The United States and Canada are working together to increase the security as well as the efficiency of the border between our two countries. We are not "tightening" the border. We are making it smarter, so that people and goods can continue to cross the border in a smooth and efficient manner. A smart and secure border is in everybody's interest and will help to promote the prosperity of both countries.

U.S. AMBASSADOR DAVID WILKINS

Learning to see particular Canadians as dangerous is a project to which new identification technologies are essential. Identifying homegrown suspects is no easy task. Biometric technologies are touted as being able to sharpen the edges of a border made soft by its long-standing characterization as an unmilitarized zone separating special friends, as well as to outsource the border through biometric screenings for Canadians before they ever leave home. In considering how to achieve this goal while keeping the border open to trade, a number of government experts decried as outdated the idea of the creation of a northern wall. They insisted instead on the importance of a *virtual* border, one that could keep out terrorist threats at the same time that it eased the passage of legitimate businesspersons, "facilitat[ing] low-risk travelers and interrupt[ing] the flow of higher-risk travelers" ("Securing an Open Society" 2004:45). Torin Monahan (2010) argues that technological surveillance is an integral part of contemporary configurations of spatial exclusion. As such, biometric technologies are essential to a virtual border, as they are able to sort bodies into desirable travelers and terrorist threats while simultaneously bringing the hard edge of the border into relief.

OUTSOURCING THE BORDER

Biometric technologies are currently identified as the most important border technology. Border agreements discursively represent biometric technologies as able to accomplish one of the key features of a post-9/11 security environment: to move the U.S.-Canada border away from North America. In a report published in 2005, the Canadian Senate Committee on National Security and Defense (2005) suggested that securing the border required that "threats" to Canadian security be identified as far away from North America as possible. Outsourcing the border refers to specific techniques that a state may use to deterritorialize its borders. The aspiration is illustrated in figure 9.

This committee's report depicts Canada as a cell that can be protected by thickening its outer wall, using new technologies to protect the sanctified inner space of the Canadian state. That is, using biometrics Canada will shift the border away from the traditional edges of the nation-state to the outside edges of North America. The committee's recognition that, along with the U.S., it must securitize the broader entity of North America is consistent with what Didier Bigo (2002) describes as states' attempts to extend their spheres of control by "delocalizing" their borders. Both Canada and the U.S. are thus attempting to "outsource" their borders. In the *Statement of Mutual Understanding on Information Sharing* between the two countries the primary way to secure the state is described:

> The best way to secure our borders is to identify and intercept persons posing security risks as early as possible, and as far away from our borders as possible.
>
> Information sharing supports the Multiple Borders Strategy, which focuses control on measures overseas, where potential violators of citizenship or immigration laws are intercepted prior to their arrival to the United States or Canada. (Canadian Department of Citizenship and Immigration, U.S. Immigration and Naturalization Service, and U.S. Department of State 2003)

Biometrics are one of the primary technologies used to achieve the outsourcing of the border. The critical legal theorist Audrey Macklin (2001:385) documents the ways that Canada pursued an international strategy of border patrol. Although the strategy is not new, it is facilitated by the use of biometric technologies. This helps to explain the

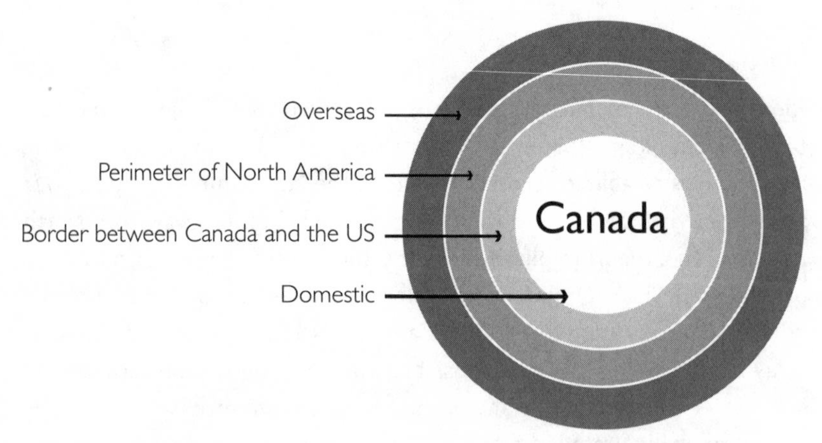

FIGURE 9 This image was taken from "Borderline Insecure," the interim report of the Canadian Senate Committee on National Security and Defense, published in June 2005. Illustration by Tracy Ellen Smith, Creative Design Resources, Inc., CDRtexas.com.

proliferation of biometric programs used to test people before they depart from their point of origin. For example, as part of Canada's $3.5 million program to screen immigrants before they arrive on Canadian soil, the government is opening an office in Singapore to collect biometric information from immigration applicants. The information will be used to determine "the identity of immigrants, although how this will be done is not clear" ("Canadian Biometric ID Documents Public Forum" 2006).

Rather than seeing the border as a one-dimensional dotted line along a surface, biometrics represent the border in 3D: "Effective border management requires governments to treat the border as more than a single line at which threats can be intercepted" (Government of Canada 2004:41). Thus the border multiplies endlessly outward, replicating itself such that no one line may be identified:

> Canada and the United States are pursuing a regional approach to migration based on the *Multiple Borders Strategy*. The *Multiple Borders Strategy* views the border not as a geo-political line but rather a continuum of checkpoints along a route of travel from the country of origin to Canada or the United States. At every checkpoint along the travel continuum—visa screening; airport check-in; points of embarkation; transit points; international airports and seaports; and the Canada–United States border—there is an opportunity for the participants to link the person and the document and any known intelligence. (Cana-

dian Department of Citizenship and Immigration, U.S. Immigration and Naturalization Service, and U.S. Department of State 2003)

The reference to intelligence suggests that the goal is not simply immigration confirmation, but terrorism intelligence, pointing to the connections being drawn between immigration and terrorism. This linking of "person and document" is accomplished through the biometric inspection.

Biometric technologies feature centrally in the numerous border initiatives signed after September 11, 2001. The major Canadian, U.S., and bilateral programs governing the post-9/11 border are the Smart Border Declaration, Canada's National Security Policy, and the Western Hemisphere Travel Initiative. These new identification technologies also are central to ongoing discussions between the United States, Mexico, and Canada concerning the possibility of a new trilateral border initiative, the Security and Prosperity Partnership of North America.

The U.S.-Canada border is not only outsourced to other nation-states to screen "risky travelers." New technologies are also deployed to outsource the U.S. border onto Canadian territory. As we saw with the introduction of NEXUS-Air, most major Canadian airports, including those in Vancouver, Edmonton, Calgary, Winnipeg, Toronto, Ottawa, and Montreal, now have U.S. customs screening located where travelers may have their identity biometrically confirmed by an iris scanner. Biometrically scanning travelers prior to their entry into the U.S. precludes the admission of those deemed dangerous rather than requiring their expulsion—a process central to outsourcing the U.S. border (Foreign Affairs and International Trade Canada 2004:6). U.S. requirements that Canadians obtain biometric passports as a result of the Western Hemisphere Travel Initiative (WHTI) will further the process of outsourcing the border. When the WHTI comes into effect, all Canadians, not only NEXUS-Air members, will be subject to biometric identification by U.S. customs officials before they leave Canada. This is part of the larger plan to internationalize the smart borders model: "Through institutions such as the G8, the World Customs Organization, the International Maritime Organization, and the Asia-Pacific Economic Co-operation Forum, we will seek to enhance international standards and to internationalize our Smart Borders programs" ("Securing an Open Society" 2004). Thus the border has become the latest North American export. Besides outsourcing the U.S. border into Canada biometrics are central to the outsourcing of a North American perimeter abroad. What formerly was understood to be just a coastline or a line on the ground between two nations is now imagined to be composed of the individual bordered bodies that make up the key information contained in the customs official's computer, telling each inspector who is in the United States or Canada, why, and for how long.

VISUALIZING THE INVISIBLE

Throughout this book I problematize claims to visibility as transparency, extending Donna Haraway's (1997) term "corporeal fetishism" to demonstrate that the process of making an object visible is fraught with multiple meanings, contradictions, and errors. Although making the U.S.-Canada border visible is presented as a simple effort to map the natural boundary between two nation-states, this project is much more than geographical, as the process of visualizing the border goes beyond the straightforward detection of the edges of each country.[3]

The International Boundary Commission is the bilateral organization that ensures that the 5000-mile border between Canada and the United States is physically demarcated. The IBC operates with an annual budget of $1.4 million from the U.S. and $2 million from Canada (Bowermaster 2007), a contentious allocation given that both countries are supposed to share equally in the funding (Porter 2007). Regardless, the IBC's budget to physically clear the border is modest in comparison to the budgets earmarked by both countries for high-tech border security. Canada alone earmarked U.S. $368 million to increase border security in 2007–2008 (Day 2007), including unmanned hover drones to patrol the border, higher-tech forms of identification for those crossing, and biometric technologies.

The IBC's goal is to ensure that even those persons who accidentally stumble across the border cannot fail to note its existence. Although this task is identified as a simple undertaking, finding the US-Canada border has proven to be no easy task. Regular markers are placed along its length, though the geography makes the border difficult to find. Rather than being located along easily identifiable natural features like rivers and mountains, the border runs through heavily forested areas, crosses the prairies and the northern tundra, and weaves a watery line through the middle of the Great Lakes. In 2006 the *Ottawa Citizen* reported, "The United States is eager to install a battery of surveillance towers, motion sensors and infrared cameras to monitor the Canada-U.S. border. Now if only they can find it" (Alberts 2006). Old maps drawn in the 1930s and thorny vegetation up to twelve feet tall provide significant challenges to border detection (Alberts 2006). Even those places that are cleared quickly become overgrown with trees, brush, and snow: "The U.S. and Canada have fallen so far behind on basic maintenance of their shared border that law enforcement officials might have to search through overgrown vegetation for markers in some places" (Associated Press 2006). This inability to find the border causes serious consternation among officials of both countries.

Border officers frequently express their apprehension about being charged with the difficult task of securing this largely invisible border. Dennis Schornack, the former U.S. commissioner of the IBC, announced, "I can send you places where you just can't find the border" (Associated Press 2006). As he said, "If you can't see the boundary, then you can't secure it" (Alberts 2006). Schornack feared a "real diplomatic dispute" between the two countries as a result. Officials responsible for guarding the border with post-9/11 fervor found themselves playing a game of hide and seek, highlighting the tension over the paradoxical process of making the border visible.[4]

It is in this post-9/11 context that new technologies come to play a complicated role in boundary identification. Part of the fraught process of uncovering the material edge between the United States and Canada, these technologies raise poorly understood ideological issues. The hard work of cutting down hedges to locate the border fails to draw the attention and excitement of funders to the IBC; technological solutions attract far greater enthusiasm. As Dennis Schornack said repeatedly, "I've talked and talked, and we don't seem to be getting anywhere" (Alberts 2006). Weed whackers are not enough in what I call the technofantastic manufacturing of the border.

Enter biometrics, a visualization technology that invites increasing attention in the post-9/11 spotlight. In detailing future strategies for U.S. security, two of the key recommendations in *The 9/11 Commission Report* are to include biometric identifiers in travel documents and develop a biometric entry-exit screening system for visitors to the United States (National Commission on Terrorist Attacks upon the United States 2004). Biometric technologies have become an essential component of border security, providing smarter, sexier, and more profitable solutions than those afforded by their analog counterparts—chain saws, axes, and manual labor.

THE SMART BORDER DECLARATION

On September 11 U.S.-Canada border traffic slowed to a crawl, including goods traveling back and forth. Given Canada's economic dependence on the U.S. after NAFTA, intense lobbying from the Canadian business community resulted in a number of measures aimed at keeping the border open to trade. The first result of these discussions was the Smart Border Declaration, which Canada and the United States rushed to sign on December 12, 2001, a scant three months after 9/11. The Smart Border Declaration is a wide-ranging border accord introduced by power players in both countries. Deputy Prime Minister John Manley, who supported the war in Iraq, was regularly described as the Canadian politician of the Liberal Chrétien government in whom the Bush administration had the greatest trust (Clarkson 2002; Barlow 2005). Given the emphasis on security and surveillance after 9/11, the choice of Tom Ridge to announce the initiative, as Ridge was soon to become the first secretary of the Department of Homeland Security, was significant and demonstrated a shift in focus from trade with Canada to security in the U.S. Biometric technologies are central to both.

The very first point of the Smart Border Declaration's thirty-point Action Plan addresses the need for the adoption of biometric technologies:

> #1 BIOMETRIC IDENTIFIERS
> The United States and Canada have agreed to develop common standards for the biometrics that we use and have also agreed to adopt interoperable and compatible technology to read these biometrics. In the interest of having cards that could be used across different modes of travel, we have agreed to use cards that are capable of storing multiple biometrics.
>
> Our two countries have also worked with the International Civil Aviation Organization (ICAO) to approve and adopt international standards for the use of biometrics in travel documents. This international cooperation allowed ICAO to announce, on May 28, 2003, that the facial recognition biometric had been selected as the globally interoperable biometric. ICAO also certified two other biometrics for secondary use

[iris recognition and fingerprints]. (Foreign Affairs and International Trade Canada 2004)

In listing biometrics as its first objective, the Smart Border Action Plan ensured that these technologies would be central to remaking—and outsourcing—the border. The Smart Border Declaration has been the subject of five joint status reports since it was signed. During this time, although the role of biometric technologies has expanded, promoting their use has remained at the top of the bilateral agenda.

Aside from NEXUS-Air and other programs designed to expedite border crossings for businesspersons and other regular travelers, the Smart Border Declaration also gave rise to a number of biometric initiatives designed to provide additional screening of persons designated as security risks. Since 9/11 the government of Canada has compelled permanent residents to obtain identity cards equipped with a chip that has the capacity to store biometric identification information. In this way biometric technologies may be deployed to subject newcomers to Canada to a secondary identity test, outsourcing the border onto individual bodies. In addition the sharing of biometric information on immigrants and refugees to both countries is provided for in the *Statement of Mutual Understanding on Information Sharing* (Canadian Department of Citizenship and Immigration, U.S. Immigration and Naturalization Service, and U.S. Department of State 2003), another agreement that arose out of the Smart Border Declaration. The title of this bilateral effort, emphasizing mutuality and understanding between Canada and the United States, reveals much about the forms of cooperation found within its pages, as the information that may be shared between the two countries includes but is not limited to the following:

- Name
- Alias(es)
- Gender
- Physical description
- Date of birth
- Country of birth
- Country of last permanent residence
- Nationality or nationalities
- Biometrics including photographs and fingerprints

- Work history
- Military service
- Links with terrorist or organized crime groups
- Citizenship or immigration enforcement status and history
- Travel carrier information
- Passport and travel document information
- Personal identification numbers
- Travel routing, itinerary, and history
- Telephone numbers
- Addresses
- Marital status and family composition
- Immigration status
- Previous immigration violations
- Outstanding immigration and criminal warrants for arrest
- Criminal history and convictions for which no pardon has been granted in Canada or the United States
- Occupational information
- Education
- Grounds of inadmissibility
- Grounds of removal
- Documents submitted in support of an application to Citizenship and Immigration Canada, U.S. Immigration and Naturalization Service, or U.S. Department of State or to their successors
- Other criminal and security intelligence

This agreement thus provides for the sharing of biometric information across borders. Additionally point 4 of the Smart Border Action Plan, Refugee/Asylum Processing, established a binational working group that meets regularly to implement the systematic exchange of personal information between the two countries. In August 2004 the working group agreed to study the feasibility of comparing biometric identifiers (fingerprints and facial recognition), in addition to comparing records based on biographical data (Foreign Affairs and International Trade Canada 2004).

This twofold use of biometric technologies to scrutinize bodies represented as high-risk at the same time as they facilitate the movements of business travelers mirrors the dual purpose of the Smart Border Declaration.

SECURING AN OPEN SOCIETY

Building on the Smart Border Declaration of 2001, in 2004 Prime Minister Paul Martin released what he described as "Canada's first-ever comprehensive statement of our National Security Policy." Laying out an action plan for preparing for and responding to threats while simultaneously maintaining Canada's "historic openness"—that is, while claiming to protect "core Canadian values of openness, diversity and respect for civil liberties"—the statement identified a number of areas key to addressing security concerns, including protecting Canada and Canadians at home and abroad, ensuring that Canada is not a base for threats to its allies, and contributing to international security. Once again, the use of biometrics to ensure the protection of the U.S.-Canada border is of paramount importance:

> Given the critical role that biometrics increasingly play in authenticating the identity of travellers, the Government will also work toward a broader use of biometrics. In accordance with international standards, Canada will examine how to use biometrics in our border and immigration systems to enhance the design and issuance processes of travel and proof-of-status documents and to validate the identity of travellers at our ports of entry. ("Securing an Open Society" 2004:45)
>
> Five new initiatives address security at the U.S.-Canada border:
>
> - Deploy facial recognition (digitized photograph) biometrics on Canadian passports
> - Complete RCMP electronic fingerprint system
> - Streamline the refugee determination process
> - Further develop next generation smart borders agenda
> - Apply smart borders principles internationally. (41)

The last point represents biometrics as helping to achieve the shared goal of exporting the border away from North America. "Effective border management requires governments to treat the border as more than a single line at which threats can be intercepted" (41). Predictably those identified as in need of biometric screening are "immigrants, refugee claimants, and visitors" (41).

U.S. scrutiny of the biometric information on Canadians who seek entrance at the border is largely a result of the Western Hemisphere Travel Initiative, signed in 2007. The WHTI is the latest in a series of security policies governing the introduction of biometrics after 9/11. Krista Boa (2006) demonstrates that while the name "might imply that the [Western Hemisphere Travel Initiative] is a multi-lateral initiative, it is not." The agreement mandates that anyone entering the United States, including U.S. citizens, must "present a passport or another type of identity and citizenship document approved by the Department of Homeland Security: the passport is the preferred document." The passport requirement is already in effect for those who wish to travel by air between the U.S. and Canada (Canadian Border Services Agency 2007). The WHTI received an enormous amount of critical attention in Canada, where the passport requirement is cause for particular concern, as checking each passport may seriously interfere with commercial traffic at the border. By revoking earlier agreements allowing Canadians to enter the U.S. without a passport, the WHTI initiative represents a concrete end to Canadian exceptionalism.

The WHTI mandates biometric travel documents for all Canadians entering the U.S. Each document presented "must establish the citizenship and identity of the bearer, and include significant security features. Ultimately, all documents used for travel to the U.S. are expected to include biometric technologies that can be used to authenticate the document and verify identity" (Moss 2005). Biometric companies have become the guarantors of border security. Although other border agreements, including the Smart Border Declaration and Canada's National Security Policy, made mention of the need to develop biometric passports, the WHTI clearly asserts that Canadians will need to have biometric identification as of June 2009. Following the announcement of the WHTI, changes to Canada's passport order requiring passport photos to be digitized suggested that the WHTI had prompted Canada to meet biometric passport requirements using facial recognition technology (Boa 2006). Even President George W. Bush was surprised by this new requirement for Canadians, reminding us of the challenges of transforming Canada into a terrorist threat. Upon being informed in 2005 that Canadians would need biometric passports to enter the United States, Bush appeared confused:

"When I first read that in the newspaper, about the need to have passports . . . I said, what's going on here. I thought there was a better way to do—to expedite legal flow of traffic and people. Evidently this has been mandated in law" (Bush, 2005).

Significance: Biometrics and the Border

Like the construction of biometric technologies themselves, and as a result of the post-9/11 reliance on biometric technologies to visualize the boundaries of the nation-state, the process of making the U.S.-Canada border visible depends upon practices of inscription, reading, and interpretation that are assumed to be transparent and self-evident and yet remain complex, ambiguous, and inherently problematic. While manufacturing the border is represented as straightforward, it is in fact a highly contested process. Attempts to engrave this line on the North American continent reveal the complex tangle of geographical, technological, national, bureaucratic, gendered, and racialized discourses at work—all of which play a role in manufacturing the border. Most important, the emphasis on the ability of biometric technologies to reify the boundary and make the border newly visible creates effects that are far from neutral. What is the significance of this transformation of the U.S.-Canada border?

PROFITABLE TECHNOLOGIES

State investments in biometric technologies for the purpose of defining the U.S.-Canada border represent one of many post-9/11 economic opportunities for the business community. Although many of the cost breakdowns are kept secret by the Canadian government, the political scientist Emmanuel Brunet-Jailly shows that the Smart Border Declaration signaled the beginning of a period of unparalleled spending on border security. The expenditures began when Prime Minister Jean Chrétien announced a budget increase of 7.7 billion Canadian dollars for security and the military a scant three months after 9/11 ("$24B Spent on Security in Canada Since 9/11" 2008). Allotting $1 billion to screening newcomers to Canada, to detention, and to administering refugee claims, the Canadian government also allocated funds to the development of new forms of identification. These included a permanent resident card and the development of fraud-resistant passports, both of which contain biometric identifiers (Brunet-Jailly 2006) and are part of the increased surveillance of Canadians. Canadian border agencies received $1.2 billion to implement the goals of the Smart Border Declaration, more than half of which ($646 million) was for the development of new technologies (Brunet-Jailly 2006). In 2004 Canadian national security policy budgeted $690 million, in addition to the $8 billion that had already been spent on security, to address other "gaps" ("Securing an Open Society" 2004:5). In 2007 Stockwell Day, the minister for public safety, announced a further $431.6 million to be invested over five years "to reinforce smart, secure borders" as part of the latest border initiative, the Security and Prosperity Partnership of North America (Day 2007). All of these costs, combined with increased military spending, are now thought to total $24 billion in Canadian expenditures on security since 9/11 ("$24B Spent on Security in Canada Since 9/11" 2008).

U.S. expenditures on post-9/11 security measures are even greater. In 2003 the national budget for security increased by 1,000 percent. Expenditures for border security rose from $2 to $11 billion (Brunet-Jailly 2006). The Immigration and Naturalization Services budget was $1.2 billion, which permitted the U.S. to double the number of border patrol and inspection agents. It also allowed for the creation of an entry-exit system that can track the arrival and departure of every visitor to the country (Brunet-Jailly 2006); that system cost $380 million. The Government Accountability Office, the nonpartisan, investigative arm of Congress, estimated that the addition of biometrics to the border would cost billions of dollars; the addition of biometrics to visas alone will cost between $1.4 billion and $2.9 billion. Predictions of recurring annual costs range from $700 million to $1.5 billion, depending on the combination of biometrics used (U.S. Government Accountability Office 2002a, 2002b; Bergstein 2003). The anticipated expense of the initial cost of adding biometrics to passports would be between $4.5 billion and $8.8 billion per year, again depending on the biometric technology used: passports with biometric fingerprinting, iris scanning, or facial recognition technology are more costly than those that contain biometric fingerprints only. In either case, the system is estimated to require $1.6 billion to $2.4 billion per year in maintenance (Morgan and Krouse 2005).[5]

The biometric industry describes 9/11 as a tremendous business opportunity. For example, the biometrics group of the Canadian Advance Technology Alliance was formed within six months of the terrorist attacks. The company's website asserts that "the events of September 11th have completely turned around the perception of the biometrics industry"

FIGURE 10 Cartoon by Khalil Bendib, originally published on http://www.corpwatch.org alongside an article about the dangers of privatizing border security and immigration control. Courtesy of Khalil Bendib.

(CATA Alliance 2007). The close links between business opportunities and biometrics are carefully documented by Kelly Gates (2004) in her excellent study of the emergence of biometric facial recognition technology. Gates demonstrates that the biometrics industry is setting the terms of the debate as to whether facial recognition technology should be adopted, whether it works efficiently, and whether the technology protects privacy. Unsurprisingly, given their vested interest in the adoption of these technologies, the biometric industry's answer to these above questions is yes, yes, yes.

Compared to the $4 million budget of the International Boundary Commission to keep the six-meter-wide border free of debris, the biometric industry's potential for profits is enormous (fig. 10).

The cost required to secure the U.S.-Canada border is so great that it seems unlikely that this model is sustainable (Brunet-Jailly 2006). Biometrics helped transform the border from a space that eases the transactions of big business to a space that *is* big business. As the science studies theorists Bruno Latour and Steve Woolgar (1979) remind us, reality may be defined as that set of statements that is too costly to give up.

Biometrics are yet another visualization tool in the arsenal of technologies aimed at making secure the transfer of mobile bodies from one nation to another. Before biometrics, passports contained "digitized photos, embossed seals, watermarks, ultraviolet and fluorescent light verification features, security laminations, microprinting and holograms" (U.S. Department of Homeland Security, Bureau of Customs and Border Inspection 2006:16). Clearly the role of biometric technologies at the border goes beyond the need to secure travel documents. Despite the long list of existing technologies to secure passports, those advocating for the addition of biometrics to U.S. and Canadian passports are successful in part because they represent biometrics as able to transfer the markers of security onto the body itself.

The industry claims that biometric technologies are able to locate the otherwise invisible border on each individual body as it attempts to cross the border through biometric "practices of looking" (Sturken and Cartwright 2001). Biometric practices of looking include identification and verification. In part, biometric technologies are useful in that they are imagined to provide border officials with mechanical objectivity, held up as the moment when the objective gaze of the scanner replaces the subjective gaze of the customs official. In this way the industry makes claims about the ability of biometric technologies both to efface subjectivity and to preclude document fraud through the unblinking screen of the biometric scanner. Mechanical biometric failures reveal that their mechanical objectivity is an illusion, a visual trick designed to hide subjectivity from view. Yet given the anxieties at the U.S.-Canada border, including concerns about newly racialized Canadian bodies and terrorist threats, biometric technologies able to accomplish these multiple security tasks while remaining scientifically neutral are doubly appealing.

What does it mean for biometrics to render each body visible at the border? In her impossibly titled book *Modest_Witness@second_Millennium.FemaleMan©_Meets_OncoMouse,* Donna Haraway (1997) reveals the dangers of images that serve to simplify complex phenomena.[6] Using the example of genetics, Haraway argues that representing genes as discrete objects such as building blocks or blueprints is a form of "genetic fetishism" (141). The project of making a "gene into a thing" is intimately bound up with capitalist enterprise, from pharmaceuticals to medical science, all of which invest in accessing the gene in an essentialized, commodified form. In the same way the development of biometric technologies is occurring at a time when the state aims to make its citizens newly visible for the purpose of governance, whether in the prison system, in the welfare system, or at the border.[7] An example of corporeal fetishism, biometric technologies make the body into a "thing," a governable entity whose compliance is inevitable. Biometric technologies usefully render "the body as a kind of accessible digital map, something easily decipherable, understandable, containable—a body that is seemingly less mysterious than the body that is popularly conceived and individually experienced" (Sturken and Cartwright 2001:302). Standardizing bodies into binary code in a process of corporeal fetishism, biometric scientists construct a simplified material body that does not acknowledge the ways that this binary map of the body reflects the cultural context in which it was developed. And yet biometric representations of the body reflect the contemporary obsession with digital representations of the body, including a contemporary aesthetics of digitization, as well as a contemporary cultural moment preoccupied with identifying suspect Canadian bodies in the name of security. Rendering these bodies as code additionally suits the needs of what the communications theorist Dan Schiller (1999) calls "digital capitalism." The flimsy material body is rendered rugged as biometric technologies make bodies replicable, transmittable, and segmentable—breaking the body down into its component parts (from retina to fingerprint) in ways that allow it to be marketed more easily in the transnational marketplace, either as a security risk or a desirable consumer. The circulation of bodies in these transnational networks of global capital is a system that Simone Browne (2009) refers to as an economy of bodies. Key to this system is the body's passive compliance with the new imaging processes so that science can manage the body better, more efficiently, and seemingly without error. Unlike the unruly material body, biometric bodies offer up a text over which scientists may have "absolute experimental control" (Haraway 1989:234). The knowledge generated by the use of biometrics to test identity is asked to perform the cultural work of stabilizing identity, conspiring in the myth that bodies are merely containers for unique, identifying information which may be seamlessly extracted and then placed into a digital database for safekeeping or which may then be circulated in the global marketplace.

IDENTIFYING SUSPECT BODIES

In her dissertation on the connections between the representations of health care policy, immigration, and the U.S.-Mexico border in American media during the 1990s, Maria Ruiz (2005) argues that anxieties about the southwestern boundary are represented through the construction of the border as a line of infection. Specifically Latina/os are regularly described either as a source of infection coming across the border to contaminate the United States or as an unstoppable epidemic of immigrants overwhelming the boundary between North and South. Discursively conceptualizing the U.S.-Mexico border through the trope of infectious disease has material ramifications (Ono and Sloop 2002). Representing Latina/os as a source of disease rather than as suffering from disease means that resources can be channeled into the surveillance of immigrants rather than into the care of their health, resulting in proposed legislation (such as Proposition 187) that attempted to deny health care to those without status in the United States.

The U.S.-Canada border is being reconceptualized through the trope of terrorism. In imagining the border as a site of terrorists and in constructing the terrorist threat to the United States through the bodies of immigrants and refugees to Canada, Americans have racialized Canadians. The construction of the border as a racialized space that must be secured invites particular solutions. In part the problem of the border comes to be understood as one that must be rectified through changes to Canada's immigration and refugee policy, accompanied by high-tech solutions able to visualize these criminalized bodies as well as outsource the border away from the vulnerable U.S. Thus vast expenditures on new technologies able to visualize newly identified Canadian threats become central to the border. Just as Ruiz shows that diseased Mexican immigrants came to stand in for the U.S.-Mexico border, terrorists—in the form of the 9/11 hijackers and the Toronto 18—are used to symbolize newly identified threats to the U.S. This helps explain the reasons for large bilateral expenditures on biometrics and their role as a ubiquitous border technology in accords signed after September 11, 2001.

Summary

One of the next major deployments of biometric technologies is at the border, as they become increasingly central to managing, and indeed manufacturing, the US's northern boundary. Although the border is regularly represented in laws and policies and by border officials as naturally embedded in the soil, they themselves embed the border using a range of technologies, a project to which biometric technologies are essential. This project of transforming the U.S.-Canada border is supported by the twin logics of corporate capitalism and racialization, as biometric companies offer technologies able to visualize "difference" in ways that will protect the U.S. from these newest invisible enemies. This is a neoliberal project that promises considerable benefits to the biometric industry. Biometrics are another in a long line of technologies claiming to be an easy technological solution to complex problems. Offering to redefine social problems as scientific ones (Harding 1993:15), biometrics portray old inequities in new ways.

In imagining immigrants and refugees to Canada through the trope of terrorism, the biometrics industry constructs the U.S.-Canada border as a place of potential threats. In racializing Canadians and Mexicanizing the border, the industry represents the border as offering up new threats in need of identification. This is a discursive representation of Canadians that has clear implications for the solutions that will be developed. Hence the introduction of biometrics as the border technology par excellence, offering new visualization technologies to definitely identify bodies and sort them according to risk—a classification process fueled by xenophobic anxieties and intimately connected to capitalist enterprise.

The border theorists Kent Ono and John Sloop (2002:5) argue, "Rhetoric shapes understandings of how the border functions and suggests that productive future work needs to be done in order to alter purposefully the meaning of borders, of nations, and of peoples." The U.S.-Canada border has been shaped by a discourse of technological neutrality and efficiency. This narrative needs to be complicated given the ways that it works to scrutinize certain bodies (immigrants and refugees) while facilitating the passage of others (business travelers) and not permitting suspect bodies to even leave their country of origin as the border is outsourced away from the edge of the U.S. and Canada. We need a new narrative that works to perform the border in ways that are not based on principles of exclusion or claims to mechanical objectivity. Instead we need policies based on principles of inclusiveness and which foster and facilitate substantive claims to equality at the border.

The failure of the International Boundary Commission to locate the border reminds us that the process of making something visible is never as straightforward as the proponents of visualization technologies, whether cartographers, doctors, or biometric industry representatives, claim it to be. Representations of biometric technologies tend to depict them working perfectly, giving credence to industry assertions that biometric technologies can reliably identify individual identities beyond the shadow of a doubt. Yet rather than operating under the aegis of mechanical objectivity, biometric technologies bring to life assumptions about bodily identity, including race, class, gender, sexuality, and disability.

five REPRESENTING BIOMETRICS

•—•—•—•—•

The truth is written on all our faces.

LIE TO ME, 2009, EPISODE 1, SEASON 1

In the Fox television drama *Lie to Me* a renowned expert in deception detection named Cal Lightman reads body language in order to help government agencies and law enforcement solve their most challenging cases. Lightman repeatedly asserts that whereas the average person tells "three lies per ten minutes [of] conversation," the body never lies. Whenever his scientific expertise is called into question by others, he effortlessly reads the body language of his naysayers in order to reveal their personal secrets, thus convincing and discomfiting his critics. Claiming to be based on the work of the scientist and founder of kinesiology Paul Ekman, whose research I explored in chapter 1, *Lie to Me* proposes that bodies are truth containers that may be compelled to reveal information whether they wish to or not.[1]

From science fiction to congressional debates, biometrics belong to an arsenal of digital technologies that are now part of the mainstay of media programming (Nakamura 2009). As I heard repeatedly at my most recent visit to a biometrics convention, no industry representative attempts to market one of his or her products without going to see the latest science fiction film featuring these new identification technologies. Biometric companies as well as government proponents of these new identification technologies claim they are indexical identifiers for human bodies. Throughout this book I have extended Donna Haraway's term *corporeal fetishism* in order to think about how biometric technologies are represented as able to dissect the body, rendering each bodily part into a separate strand of binary code and, in doing so, imbuing each with use value. Biometric representations of the body are also part of the visual culture of the War on Terror, an attempt to render the human body fully visualizable. Representations of biometrics have as much of an impact on the growth and expansion of biometrics as do technological developments in biometric science. Biometric discourse produces new forms of *scopophilia,* pleasure in looking. I coin the term surveillant scopophilia to show that the new forms of pleasure in looking produced by biometric technologies are tied both to the violent dismembering of bodies marked by racialized, gendered, classed, and sexualized identities and to pleasure in having anxieties about security resolved by biometric surveillance.

Visualizing Biometrics

In 2008 Juan Diego Catholic High School in Utah became the first school in the state to biometrically fingerprint its students. In response to parents' concerns that the school was violating their children's privacy, the systems administrator told parents they had nothing to fear: "It's just like the cop shows you see where it puts the points on the finger and we only store those points" (Cutler 2008).

Media representations of biometric technologies play a significant role in their adoption in real-life programs. From the films *Minority Report* and *The Incredibles* to the TV series *24* and *Battlestar Galactica,* biometric technologies are a staple of contemporary media. As biometrics become central to media texts, so too do new biometric "practices of looking," to borrow Sturken's and Cartwright's (2001) phrase to describe the active nature of looking and how it constitutes and negotiates complex power relationships. One of these practices is surveillant scopophilia. In her essay "Visual Pleasure and Narrative Cinema," Laura Mulvey (1978:64) argues that women in film symbolize the "threat of castration and thus unpleasure." Mulvey explains that film narratives commonly construct tension around the representation of the female object. This tension may be resolved in a number of ways, including through the "devaluation, punishment or saving of the guilty object" (64). One of the ways that pleasure is cinematically produced and the spectator's anxieties allayed is through the filmic dismemberment of women's bodies. Biometric representations of the body too are a form of "digital dissection" (Amoore and

Hall 2009). Often they are cinematically used to resolve anxieties produced by the film's narrative around security—security that is threatened by othered bodies. Using biometric technologies as a form of surveillance of suspect bodies and then reducing those suspect bodies into their component parts helps to allay the anxieties created around security through surveillance and dismemberment.

As I noted in the introduction, although Steven Spielberg's film *Minority Report* is not the first movie to use biometrics, the film's release in 2002 shortly after 9/11, at the same time as the explosion in the state's use of these technologies, has made it one of the key sites for the production of meaning about biometrics. *Minority Report* is set in Washington, D.C. in the year 2054, where it is impossible to move from one location to another without being identified. Biometric technologies are central to this process, achieved through continual eye scans. From stores to billboards to the entrance to the subway, biometric scanners are ubiquitous. In addition Washington in 2054 has implemented a Department of Precrime. Relying on the superpowers of three precognitives capable of visualizing future murders, the city has become nearly crime-free. John Anderton is the head of Precrime, the unit responsible for responding to the precognitives' visions and stopping murders before they occur.

The turning point in the film occurs when the precognitives identify Anderton himself as the murderer of a man whom he has never met. In a famous sequence, Anderton flees the department but finds he has nowhere to hide due to the omnipresent eye scanners. He is continuously identified by a series of smart billboards inviting him to enjoy a range of products, from a trip in the Caribbean to a pint of Guinness. In an effort to hide his identity he has his eyes surgically replaced with those of another man. In one of the film's most gruesome scenes—one which references contemporary fears about the nightmarish possibilities of biometric identification—Anderton is forced to allow state robots to examine his newly inserted eyes before they have had the chance to fully heal.

Minority Report is regularly criticized by the biometrics industry as presenting a falsely dystopian view of biometric technologies and their implications for surveillance. In a presentation titled "From Hollywood to the Real World" one industry representative, Catherine Tilton (2004), bemoans the representations of biometric technologies in the media. Suggesting that these depictions produce an "Orwellian effect," Tilton asserts that portraits of these new identification technologies unfairly suggest that they invade our privacy, are easy to spoof, or remain in the land of science fiction. Arguing that these representations of biometrics are myths and harm actual, informed understandings of these technologies, Tilton is representative of the industry's frustration with *Minority Report,* which remains linked in the public imaginary with biometric technologies. As one official told me during a conversation about his company's biometric billboards, companies need to be careful "to assuage the worries of consumer-advocacy watchdogs and their braying quotes of 'Minority Report' " (Paolo Prandoni, personal communication, 2008).

Although *Minority Report* purports to be a dystopian representation of biometrics, its critique of these technologies is limited. In depicting biometric technologies as ubiquitous, the film naturalizes their expansion. Although they can be fooled by eye replacement surgery, *Minority Report*'s biometric scanners otherwise function perfectly, suggesting that only extreme and violent bodily reconfiguration is sufficient to trick this technology.[2] The well-functioning identification technologies act as visual arguments for their adoption, persuading politicians and consumers alike to accept their implementation in multiple locations.

For example, though the biometric industry rejects *Minority Report* and fears its impact on public understandings of biometric technologies, congressional representatives turn to this film for inspiration. In one congressional hearing some debated whether it would be possible for the U.S. government to use these technologies to create a precrime division just like the one found in the movie, able to "head off crimes or attacks, not just investigate them after the fact": "Sort of predicting, something out of the movie Minority Report? How would you head off a crime? How do you identify a potential crime or criminal? You have predictive alga rhythms [*sic*] or profiling, risk scoring? It seems fascinating as a former prosecutor; could you just put us all out of, out of business? Can you tell who's going to commit a crime?" (Specter 2005).

The blurred boundary between science fiction and reality is not new. As David Lyon (2007) shows, the inspiration for electronic tagging originally came from *Spiderman.* Science fiction films inspire the use of these technologies for systems of "actuarial justice," in which information is collected and warehoused by the state before it is needed (Feeley and Simon 1994). As Congress debates the possibilities of using biometric technologies to administer actuarial justice in the form of real-life pre-crime departments, we see the impact of the representations of these new identification technologies.

It is worth noting that although *Minority Report* does naturalize biometric functioning, the film also suggests a useful critique of the interpretation of scientific images as truth. In an interesting sequence a representative from the Justice Department, Danny Witwer, comes to speak to Anderton about the role of the Precrime Department.

> WITWER: Science has stolen most of our miracles. . . . I find it interesting that some people have begun to deify the precogs.
> ANDERTON: Precogs are pattern recognition filters. That's all.
> WITWER: Yet you call this room the temple.
> ANDERTON: Just a nickname.
> WITWER: The oracle isn't where the power is anyway. The power has always been with the priests.

Witwer is suspicious of the Precrime Department and highlights the importance of Anderton's analysis of the precognitives' visions. Rather than suggesting that the department's images are handed down to them from either the lab or the heavens, free from human interpretation, the film highlights the power of those "priests" who produce the dominant meanings of a particular set of facts.

> ANDERTON: Cut the cute act, Danny boy, and tell me exactly what it is you are looking for.
> WITWER: Flaws.
> ANDERTON: There hasn't been a murder in six years. . . . There is nothing wrong with the system. It is perfect . . .
> WITWER (simultaneously): Perfect. I agree. But there's a flaw. It's human. It always is.

Emphasizing that human hands construct the precognitives' narratives, the film opens the door to understanding scientific practice as an interpretive act. In this way *Minority Report* suggests that the priests of biometric science, whether industry insiders or government representatives, have a powerful role in rendering biometrics as able to reveal bodily truths, an assertion that we have seen guarantees some bodies inclusion, whereas others are slated for biometric exclusion.

The Visual Culture of Biometrics Technologies

In the science fiction television series *Dark Angel*, a secret government project, Manticore, has been conducting experiments aimed at producing supersoldiers. Mixing both animal and human DNA in special cocktails, each of the Manticore progeny is born with a barcode on the back of its neck. Max, known by the number x5 452, is the ass-kicking grrl heroine of *Dark Angel*. She and eleven other supersoldiers escaped from Manticore's training facility when they were still children. On the lam, Max and her transgenic siblings spend much of season 1 trying to avoid recapture. One constant source of anxiety to them is their barcodes. The U.S. surveillance infrastructure is in overdrive, and hover drones, surveillance helicopters equipped with biometric recognition technology, are constantly overhead. At one point Max tracks down one of the other escapees by following a lead about someone having a barcode removed in a tattoo parlor. Max is asked by her partner in crime why she never tried to have her own barcode removed:

LOGAN: You might want to think about having your barcode removed too.

MAX: I tried once, it feels like someone's poured acid on your skin after it's been sandblasted. It came back after a couple of weeks. It's etched into our genetic code.

LOGAN: The mark of Cain.

Such representations of biometrics show them working perfectly, telling stories of unique bodily identities that, once located, reliably give "the ultimate proof that you really are you, every time it counts" (Precise Biometrics 2006). Meant to highlight the spectacular nature of biometric recognition in contemporary visual culture, what I term the technofantastic elements of biometric representation include the moment of the biometric match, the violent dismemberment of the body to fool the biometric scanner, and the representation of biometrics as scientific, objective, race- and gender-neutral technologies. The visual representation of these technologies is also a form of surveillant scopophilia, in which we take pleasure in the identification of the biometrified body.

The moment of biometric match is ubiquitous in the representation of these new identification technologies. In her recent work on HIV/AIDS Paula Treichler notes that 9/11, an event that claimed fewer than three thousand lives, occupies far more attention and space in the American media than the AIDS epidemic, which has claimed approximately 25 million lives (Ho and Bieniasz 2008). Treichler attributes the continued importance of 9/11 in the U.S. national imaginary in part to the compelling nature of what she terms the sightbyte,[3] the collapse of the Twin Towers. The image is often described as a scene from a blockbuster disaster movie brought to life. For biometric technologies, the sightbyte is the instant of the biometric match, the satisfying point at which the computer locates an individual's biometrics, yielding the "eureka" moment that Lisa Nakamura (2009) calls the biometric money shot. This aesthetic, scientific biometric image provides a much more compelling and succinct sightbyte than, for example, a welfare caseworker flipping through the photos of the faces of people living in poverty. Like the biometric image itself, a satisfyingly contained series of zeroes and ones, the biometric match seems to guarantee our security—closing up security holes with a satisfying clink—serving simultaneously to soothe and to identify. Here surveillant scopophilia is a visual process that serves to soothe our anxieties about security, as these visual representations suggest that those bodies that threaten our security will be reliably identified.

Biometric mismatches occur with much less frequency in representations of these new identification technologies. However, when they do occur, they are usually centered on a different component of the technofantastic, the moment of the violent reconfiguration of the body in order to enable biometric identification. A staple of biometric visual culture, this scene has made an appearance in a wide variety of media, including *Dark Angel*, in which the chief villain uses a spoon to remove the eyeball of a coworker so that he can gain access to his old place of work, as well as the films *6th Day* and *Demolition Man*. We also saw this moment in the surgical removal of John Anderton's eyes in *Minority Report*. Here the biometric sightbyte is the violated, dismembered body. It is a narrative that underscores the permeable boundary between science fiction and fact. As we saw in the introduction, real-life examples of the ontological reordering of the body through surgical replacement include the migrant workers who removed their fingerprints to avoid biometric identification and the thieves who cut off someone's finger in order to start a stolen car equipped with a biometric fingerprint scanner. Thus this element of the technofantastic works to demonstrate the mutually constitutive relationship between science fiction and real life.

Although violence is highlighted in these spectacular moments of biometric recognition, in general depictions of biometric functioning elide or conceal the violence of identification. A number of visual techniques are used to anesthetize the moment of identification. Rather than the messy ink stamps which characterize movies depicting Nazis branding the passports of Jews attempting to flee the Third Reich,[4] the moment of biometric identification is often bathed in an aesthetic green or blue light, seeming to suggest a scientific, clean moment of technological identification.

Biometric technologies are naturalized as working perfectly, better than outdated, biased identification technologies like photographs and fingerprints. Key to this narrative is the claim that biometric technologies are objective or identity-neutral, able to objectively identify racialized bodies that are impossibly alike.

Representing Biometrics and Othered Bodies

Two photographs are useful for interrogating representations of biometrics as objective, race- and gender-neutral technologies. These images help to make visible the context within which biometric technologies were developed and in which they are deployed. They also explain why biometrics fail in the ways they do.

In the first photograph, a protester at an immigration rally holds a sign up to his face (fig. 11). By making reference to a Latino rather than an Arab or Muslim American identity, the protester is attempting to distinguish himself from the *real* "terrorist threat." He critically interrogates the fact that after 9/11 the bodies of all people of color in the U.S. have been rendered suspect. The sign he carries also questions the way that debates on immigration are collapsed into discussions of terrorism, security, and September 11. The sign adopts the "divide and conquer" construction of post-9/11 U.S. security rhetoric. In saying "I'm Juan not Saddam," the protester rejects bridge-building between people of color persecuted by discourses of national security. What is most interesting for the purpose of my analysis is the way that this sign references the continual need for new technologies able to uncover the sharp edges of blurred bodies. In highlighting white supremacist narratives that insist that all people of color inhabit suspect bodies, the photograph references the proliferation of racial profiling after 9/11, including cases "in which Hispanic American pilots mistaken for Arab Muslims forcibly were removed from some commercial flights by panicky crews in the days immediately after 9/11.

FIGURE 11 This photograph was found on a site hosting images of an immigration protest in New York City (www.flickr.com). Courtesy of Steven Jones.

Even a sunburned American of Norwegian and Italian descent, and a space lab worker of Jordanian descent with Pentagon security clearance, received the heave-ho, according to passenger complaints filed with the Department of Transportation" (E. Smith 2004). Demonstrating the unstable nature of racialization, these cases of the racist blurring of people of color make evident the ongoing quest for new technologies able to definitively visualize difference.[5]

The second photograph comments specifically on the unsatisfactory and outdated nature of old technologies of visualization such as photography, another recurring theme in depictions of biometric technologies (fig. 12). The image makes a visual argument for the need for new technologies like biometrics to address "indistinguishable" Orientalized bodies. A photograph within a photograph, the visible photographer is taking the photo of a group of women wearing the niqab. This image was circulated on the Internet as a joke with no caption. We are cued to understand ourselves as outside this tableau by the fact that we occupy a position

FIGURE 12 This image was circulated on a listserv as a joke with no caption. It can also be found on a number of blogs tagged as "funny" under the headline "What's the Point?" (http://onemansblog.com; http://dontvisitthis.com).

FIGURE 13 Cartoon by David Austin published in the *Guardian*, October 30, 2001. Courtesy of the David Austin estate and the British Cartoon Archive, University of Kent.

beyond even the photographer as an invisible observer, helping us to define this picture as more than someone's personal memento of a vacation. One interpretation of this photograph as an (undoubtedly racist) joke was explained to me this way: "What is the point of taking this photograph if you can't tell the bodies of these women apart?"

Another "joke" that references anxieties about the difficulty of identifying Muslim bodies is dramatized in a cartoon published on the website of the biometric scientist John Daugman (fig. 13). As we saw in chapter 2, Daugman is credited with the invention of the algorithms associated with biometric iris scanning. Clearly the scientists involved in building biometric technologies are not immune to cultural anxieties concerning the post-9/11 need for biometric technologies.

Such images serve as visual arguments for the fact that photographs, formerly the gold standard of identification technology, are no longer sufficient to identify othered bodies. Both the photograph and the cartoon also reference concerns about the imagined ramifications of this othered inscrutability or the problematic suggestion that alien bodies "all look alike." Similar anxieties concerning the limits of photographic identification underlay press coverage following the escape of the murder suspect Mustaf Jama. Jama's alleged getaway, in which he purportedly fled the U.K. by dressing in a niqab and using his sister's passport, led to assertions of the need for more robust technologies than photo identification. In that case the industry claimed that biometric technologies would have averted Jama's escape (Brady 2006). The film theorist Pat Gill (1997:164, 171) calls this claim—that new technologies can revise the past to redetermine the present—technostalgia. Kelly Gates (2004:58) documents that the magical ability of biometrics to avert disaster is repeatedly used to justify calls for their expansion.

Biometric companies and their proponents regularly represent biometrics as a needed improvement upon older identification technologies. This is apparent in a media release issued by the Biometric Consortium, one of the key centers of research, testing, and marketing of biometric technologies:

> Biometrics can be defined as an automated method of verifying or recognizing the identity of a living person based upon a physiological or behavioral characteristic; that is, it's based upon something we are or something we do.

> The word "automated" is necessary in the definition because we want to avoid the inclusion of very common, *but significantly less reliable*, methods of identification such as a photograph. (Hays 1996, emphasis added)

No longer forced to distinguish what were previously understood as identical racialized bodies, biometric technologies are represented as able to identify our true identity. Not restricted to methods of identification based only on what we look like, biometrics are able to plumb the depths of who we are and what we do. Against these problematic narratives of racial blurring, the industry claims that biometrics are able to definitely tell the bodies of people of color apart.

Racist Representations of the Body

Biometric technologies compel the state to look behind the veil during the intensified policing of Arab and Muslim bodies: Arab American, Arab Canadian, Muslim American, Muslim Canadian, and all other bodies that fall into those categories. After 9/11 the emphasis on the ability of biometric technologies to secure the nation was in keeping with the media's fixation on the niqab and the hijab: Should the state permit Muslim women and girls to cover themselves in public? Should girls wearing the hijab be allowed to compete in tae kwon do matches (Marotte 2007)?[6] Should women wearing the niqab be permitted to vote (Tu, Peritz, and Marotte 2007)? In these reports the veil represents the threatening nature of Islam while simultaneously symbolizing the imaginary divide between oppressed Muslim women and emancipated Western women.

In 2004 the International Civil Aviation Organization (ICAO) decided that facial recognition would be used as the globally interoperable biometric on all international travel documents, including passports. The ICAO is part of the UN and is the global forum that decides all international security and regulatory requirements for aviation. Facial recognition technology was chosen in part because it allows customs officers in countries without biometric technology to continue to use their eyes to determine whether travelers match their passport photo, something that would not have been possible had biometric fingerprinting or iris scanning technology been adopted. As well, renewed media attention to the risks posed by veiled terrorist bodies and female suicide bombers likely underlies the new ICAO biometric standard requirement for a face-based passport check. The result is that new biometric standards of security compel Muslim women to remove their veil for identification. In forcing state officers to look beneath the veil, the ICAO is deploying biometric technologies to invade the bodily privacy of Muslim women who cover. Bodies thus become see-through containers from which biometric science extracts identifying information as part of an "aesthetics of transparency," in which visibility ensures state security (Hall 2009). Thus biometrics render the bodies of citizen-subjects transparent, such that there are no longer any places hidden from the eye of the state.

Biometric Iris Scanning and "the Afghan Girl"

Part the veil on our killing sun.

SARAH MCLACHLAN, "WORLD ON FIRE"

The identification of one Muslim woman in particular, Sharbat Gula, demonstrates the ways that these new identification technologies are depicted as able to allow institutions to cast light (by literally and symbolically "parting the veil") into dangerous places formerly hidden from the state. The industry regularly refers to her case as epitomizing the usefulness of biometric technologies. Sharbat Gula's case is particularly salient since she is one of the most high-profile people to be identified by biometric technologies. As the unknown "Afghan girl" Gula was internationally famous, and using biometric technologies to identify her was incredibly productive for the industry's claims about the possibilities for biometrics. The biometric identification of Sharbat Gula is ubiquitous in biometric marketing materials (Das 2010; Gregory and Simon 2008) and was central to getting biometric technologies international press coverage.

In 1984 the photographer Steve McCurry accompanied the journalist Debra Denker to Afghanistan to document the plight of Afghani civilians living under Soviet occupation. Published in *National Geographic* in 1985, their article was accompanied by a photograph that has since been included in the magazine's top 100 photographs and earned McCurry the prestigious Robert Capa Gold Medal for the best photographic reporting from abroad (fig. 14).

McCurry asserts that this is one of the most widely reproduced photos in the world, found everywhere from cups to posters to rugs to tattoos. Although the image was famous, little was known about the subject herself. Usually referred to as "the Afghan girl," the young woman's image

FIGURE 14 This image of "the Afghan girl" was originally published in *National Geographic* in 1985 in an article titled "Along Afghanistan's War-torn Frontier." Photo by Steve McCurry. Courtesy of Magnum Photos.

regularly stands in for the plight of the unidentified refugees whose stories tell us not of the violence produced by decades of occupation but, as described by a journalist interviewing McCurry in 2007, "humanity's endurance" in the face of adversity (Liddell 2007). McCurry acknowledged that part of the reason the photograph attracted attention was the way the young woman's blue eyes usefully condensed the narrative of Afghani refugees into a story of "just like us" rather than a tale of "them over there":

> To most Westerners, Afghanistan is the end of the Earth, a dangerous, mysterious place where they assume the people to be completely alien. Yet McCurry's picture shows a girl who, with a change of clothes, could

effortlessly blend into any Western society. Surely, this is also part of this image's appeal for Westerners?

"Well, that's a good point," McCurry responds. "Afghanistan, being the crossroads of culture, there's a segment of the population that came down with Genghis Khan, the Hazaras. Then you have a very strong Indian influence. Then some of these people, you would swear they were Western European. If they were walking down the street in Munich, Paris or Milan, they would look like the local people. There's an interesting mix of cultures in Afghanistan and I think her face and her look illustrates that. It's not an identifiable look. It's a sort of a mix. She looks kind of Western and she looks kind of Afghan." (Liddell 2007)

The photograph of the Afghan girl thus provides exactly the right blend of consumable difference. Her image does not have the perils of the boring sameness that McCurry intimates is characteristic of the form of "inauthentic" multiculturalism found in the West: "There is an amazing mix of cultures in New York, but it's completely different. . . . In New York people are trying to assimilate, and to some extent conform to the laws and the customs and the traditions. You simply don't get the same kind of raw, original culture that you get on the streets of Calcutta or Kabul" (Liddell 2007). This photo manages to tell a narrative of difference *chez nous* that is spiced up with enough raw otherness to be compelling. In this way the photograph of the Afghan girl is the perfect disaster photo, providing exactly the right degree of edible spectacle.[7]

In 2002 *National Geographic* sent McCurry back to Afghanistan to find out more about the woman made famous by his photograph. This quest to find, unveil, and identify the Afghan girl needs to be contextualized against the backdrop of the war in Afghanistan, begun in 2001 by the U.S. with heavy participation from Canada. It was marketed as a war in the name of women's rights; in fact American and Canadian media portrayed women's liberation as needing to be unveiled in Afghanistan, and burqa-clad Afghani women became the primary signifier that women's rights were being held hostage (Ansari 2008; Faludi 2007). As Minoo Moallem (2005) demonstrates, signifying practices that locate the veil as the primary symptom of women's oppression have an imperial logic; in this case the veil was used to signify the need for outside intervention in Afghanistan. This is not a new strategy; the U.S. employed a similar logic to "unveil" Persia in order to obtain access to its petroleum (49). As women's bodies are regularly used to represent the nation, and as "more than analogy links the imperialist project of colonizing other lands and people with the phantasm of appropriation of the veiled, exotic female" (Alloula 1986:xvi), representations of the Afghan girl deserve special scrutiny. In particular *National Geographic*'s rendition of the possession and exposure of this most famous Afghani woman must be understood as part of the imperial project of bringing the light of progress to Afghanistan—with the Afghan girl's body standing in for the nation. Penetrating the space behind her veil was portrayed as being in the interest of progress, using cutting-edge technology to shed light on the spaces of premodern darkness.

As Suvendrini Perera (2003) documents, the story of locating this woman and verifying her identity is replete with colonial imagery, from tales of a duplicitous native informant leading the *National Geographic* team astray in their search to the devoted brother tracking down his sister for McCurry so that she might again be photographed. The quest itself consists in part of a Western naming ritual, in which the nameless Afghan girl is allowed to take possession of her own name. The *National Geographic* article about the search begins, "Names have power, so let us speak of hers. Her name is Sharbat Gula, and she is Pashtun, that most warlike of Afghan tribes" (Newman 2002). Part of the excitement of the discovery of Sharbat Gula was that McCurry was able to tell her that her face was famous, for she had "no idea her face had become an icon" (D. Braun 2003). *National Geographic*'s search for Sharbat Gula references a narrative of imperial nostalgia as the Afghan girl stands in for an earlier, more ignorant, primitive, warlike time, a phenomenon Malek Alloula (1986:4) describes as "a racism and a xenophobia titillated by the nostalgia of the colonial empire." This description of Gula additionally references what Minoo Moallem (2005) calls the representation of Muslim women as Western "modernity's other." In her study of Islamic fundamentalism and Iranian identity, Moallem shows that the representational practices of othered bodies are always already bound up with modern histories of race, gender, religion, and nation. We can see this in the repeated descriptions of Gula as an uneducated ("She can write her name, but cannot read"), traditional, religious Muslim woman, unable to comprehend the impact of her photographic image: "She cannot under-

FIGURE 15 Sharbat Gula in 2002 holding the photograph of her as a girl in 1985. Photo by Steve McCurry. Courtesy of Magnum Photos.

stand how her picture has touched so many. She does not know the power of those eyes" (Newman 2002).

The representation of Gula as modernity's other was accomplished in large part through a focus on her veiled figure (fig. 15). Even though the veil is not a stable signifier and historically has been used to stand in for both repression and revolution (Alloula 1986), it regularly is used to represent Muslim women as the "ultimate victim[s] of a timeless patriarchy defined by the barbarism of the Islamic religion, which is in need of civilizing" (Moallem 2005:20). As Ella Shohat and Randy Martin (2002:1) demonstrate, in Afghanistan women " 'gain[ed] face' by removing their veil, in what [was] seen as a triumph of Western modernization." Religious prohibitions that might have prevented Gula from having her photograph taken were overcome by the use of *National Geographic* funds to pay her to remove her burqa so that her face might be photographed (fig. 16; Newman 2002).

In his study of early twentieth-century French postcards of Algerian women taken during France's occupation of Algeria, Malek Alloula (1986:7) argues that the veil traditionally has served as a "powerful prod to the photographer" to discover what lies beneath and expose it to the light. It is interesting to note that once McCurry and his team located a woman they thought might be the subject of his photograph, they chose not to rely on her account or on the accounts of others who knew her to confirm her identity. Rather her identity as the original Afghan girl was confirmed for *National Geographic* by a series of experts flown over from Europe and North America (fig. 17; Perera 2003).

It is at this point that biometrics become central to the story. The Enlightenment project of scientific rationality is achieved by bringing the clarity and vision of biometric science into the darkened world of the uncivilized other, as Gula is compelled to remove her veil and step into the light and possibility of scientific knowledge. Enlightenment philosophy posits a modern "us" versus a premodern "them" as reason (captured by the science of identification) is used to trump religion (signified by the veil), as biometric iris scanning is deemed an essential component of the process of confirming Gula's identity. John Daugman is the scientist who identified the mathematical algorithms upon which all iris-scanning technologies depend, and so he was deemed sufficiently expert to identify Gula in her unlikely role as prodigal daughter. Daugman used the technology to compare the eyes in the original photograph with the eyes of the woman claiming to be the Afghan girl.

In this example of surveillant scopophilia biometrics are represented as able to tame the subject, bringing her into the laboratory for verification while promoting and providing product placement for Iridian, a biometrics company that markets iris scanning. Though anxieties are produced by the narrative of the unknown subject, difficult to find, we may be comforted that there is no subject we cannot track down and identify scientifically. Although Daugman did not have Gula's original biometric information with which to compare his current reading, since biometric iris scanning did not exist at the time that the original photograph was taken, he is still able to update old photographic technology with new biometric iris-scanning technology. Bearing many signifiers of a

FIGURE 16 Photograph of Sharbat Gula, published on the *National Geographic* website in 2002 with the caption "Portrait of a Survivor." Photo by Steve McCurry. Courtesy of Magnum Photos.

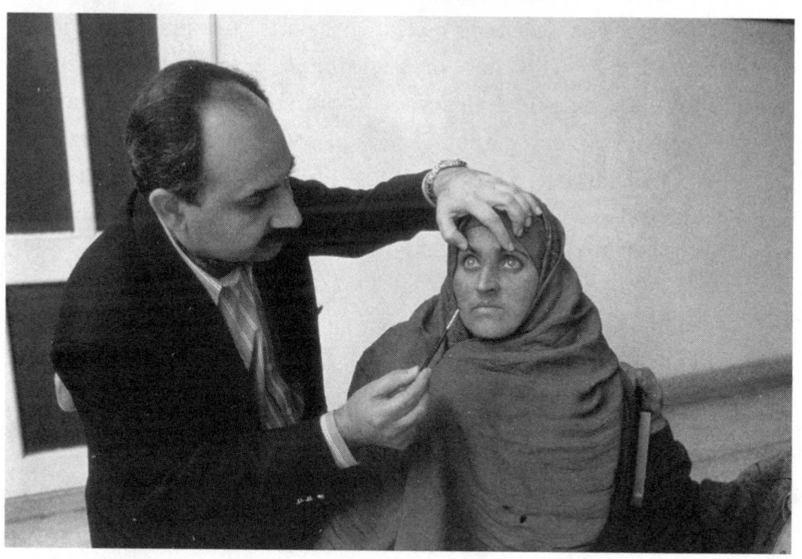

FIGURE 17 Identifying Sharbat Gula. Photo by Steve McCurry. Courtesy of Magnum Photos.

Heart of Darkness narrative by which scientific light is used to shed light on the eyes of deepest Afghanistan, the quest to find the Afghan girl and verify her identity using Western expertise usefully brings together a number of the themes that I have identified in this book: the role of biometrics in identifying Orientalized bodies, the position of biometric technologies in the profit-oriented security-industrial complex, and the deployment of biometric technologies to identify othered bodies.

In an email correspondence with the biometrics expert Henry Boitel,[8] Daugman explains that the iris-scanning identification from photograph to individual was reliable only because Gula has light-colored eyes. Thus the technology works differently on racialized bodies:

BOITEL: As I recall it, you said that iris recognition from general purpose color photographs may be possible for people with blue or light colored irises (as with the Afghan lady), but is either impossible or unlikely when such photographs depict persons with dark eyes.

Have I phrased it correctly?

DAUGMAN: Yes, that is my view.

BOITEL: Is there a reference, on the point, you can give me from your work or the work of others?

DAUGMAN: I am not aware of a systematic technical study on trying to perform iris recognition from color (i.e. visible wavelength) photographs. So, I guess the status of my pessimistic conjecture is "anecdotal!"

But I think one may appeal to a fairly universal experience: namely, the visual system of every human uses (by definition) "visible wavelengths" in the 400nm–700nm band. We all often have the experience of seeing another person with "very dark brown eyes" (e.g. a Malaysian person, or those from many South-East Asian countries) in which one cannot even detect where the pupil ends and the iris begins. The iris just looks like one big pupil, surrounded by white sclera. This common (universal?) experience is, I guess, the photonic basis for my conjecture that the visible waveband (the one used in photography) would be useless for great proportions (majority?) of human eyes.

A more technical point is this: the albedo of the iris is quite different in the visible band than in the near infrared (> 700nm). Albedo is a surface property referring to the fraction of incident

> light that it reflects (scatters) back; so albedo corresponds roughly to grey level; perfectly white paper would have an albedo approaching 100%; perfectly black matte paper 0%. Regrettably :-) the cornea lies in front of the iris, and it is a curved moist "mirror" that causes reflections of the ambient environment to cover the iris. These can be greater in magnitude than the brightness of a low albedo iris; in other words, the iris image is dominated by corneal reflections, if wavelengths are used in which the iris has low albedo. But with careful imaging in the IR band, using filters to ensure that only the IR wavelengths used to illuminate the eye are also the wavelengths allowed back into the camera, one can effectively "scrub out" the corneal reflections from the ambient environment. Obviously this tactic is impaired if the iris has low albedo; images of such eyes in photographs often appear to be dominated by corneal reflections.[9]

Gula's blue eyes are thus useful (and marketable) in two ways: first, to make her photograph compelling to a Western audience, and second, to allow her identity to be confirmed by biometric technology. In Daugman's description of her biometric identification, assumptions about the difficulty in identifying dark eyes—a problem that references assumptions concerning the inscrutability of Orientalized bodies—are identified as a "universal" experience as stereotypes concerning othered bodies are codified in the biometric iris scanner. In confirming that he has consistently found that the technologies perform differently on brown and black eyes than on blue, green, or hazel eyes, Daugman demonstrates that biometric iris scanning depends on racialization in problematic ways.

Summary

Representations of biometrics have high stakes. Revealing the permeable membrane between science fiction and real life, references by policymakers to depictions of biometric technologies in film and television demonstrate that they influence the adoption of these technologies by state institutions. From the biometric match to depictions of biometrics as able to definitively identify othered bodies, portraits of these new identification technologies rely upon scientific signifiers to argue that the process of biometric identification is objective. But the supposed neutrality of biometric recognition is revealed to be a myth. Certainly that state institutions rely upon descriptions of biometrics as able to compel bodies to speak the truth of their identities, thus compelling Muslim women who cover to remove the niqab, calls into question the objectivity of these technologies. Thus biometric technologies again fail to work in the race- and gender-neutral ways that industry would have us believe.

Representations of biometric technologies are part of the aesthetics of transparency that Rachel Hall (2009) says dominates post-9/11 depictions of security. That is, biometric technologies are represented as able to shed light into the darkness behind the veil, shining the light of the state on spaces formerly imagined to be beyond its reach. In the imperial overtones of the portrait of biometric technologies in the identification of Sharbat Gula, the subjectivity of biometrics manifests itself in their differential functioning on the intersection of racialized and gendered bodies. As these technologies are specifically deployed in order to identify suspect bodies, the impact of technological failure manifests itself most consistently on the "deviant" bodies of othered communities. That is, biometrics fail most often and most spectacularly at the very task they claim to accomplish.

conclusion **BIOMETRIC FAILURE AND BEYOND**

In the world of security and intelligence, there is no cost-benefit analysis.

WESLEY WARK, QUOTED IN "$24B SPENT ON SECURITY IN CANADA SINCE 9/11"

There are costs to clarity at all costs.

ERIC EISENSBERG, "BUILDING A MYSTERY: TOWARD A NEW THEORY OF COMMUNICATION AND IDENTITY"

The inclusion of biometric technologies in a wide range of state-funded and commercial applications continues to grow. In an interview with John Walsh of the television program *America's Most Wanted,* President Obama agreed that a DNA database containing every person's biometric information could certainly prove useful (Kravets 2010). As a result of an initiative called Secure Communities, immigrants and refugees taken into police custody now have their biometric fingerprints run through a database to determine whether they can be deported under federal immigration laws (Aguilar 2010). Google's CEO, Eric Schmidt, revealed that the company has not ruled out adding facial recognition to its search technology ("Google Refuses to Rule Out Face Recognition Technology Despite Privacy Rows" 2010). The company videosurf.com already uses computer vision technologies to search online videos, allowing users to enter a particular face as their search term.

This expansion of biometric technologies occurs despite news of their failures. Thus Obama asserts that a U.S. database of biometric information might prove useful when it became clear the U.S. watch list was "ensnared by error" (McIntire 2010). After running a photo of Osama bin Laden on the "wanted" section of their website and announcing a reward of $25 million for his capture, the FBI revealed that the sophisticated computer vision technologies they used to "age" the face of bin Laden relied upon old photos of a Spanish politician downloaded from the Internet (Govan 2010a, 2010b). Google announced that it was debating the addition of facial recognition technology a scant two weeks before it was compelled to reveal a major security breach, as its search technologies inadvertently collected and stored payload data—including significant personal information—from unencrypted wifi networks (Stroz Friedberg 2010). A report on the Secure Communities initiative collecting biometric information from newcomers to the U.S. in order to facilitate their being deported made it clear that approximately 90 percent (100,000 of 111,000) of those arrested were charged with minor infractions, and that the Secure Communities initiative had further opened the door to racist police officers trumping up charges and arresting individuals in the hopes that they would be deported (Immigration Policy Center, American Immigration Council 2009). These examples highlight many of the themes examined in *When Biometrics Fail.* We see the technological codification of existing stereotypes that imagine some bodies to be impossibly alike, as the FBI suggests that one racialized body might be switched for another in order to perfect a match. We see biometric technologies being deployed in ways that intensify existing inequalities, as a biometric database results in the increased racial profiling of immigrants and refugees and worsens their mistreatment by the police, deepening links between immigration policy makers, prisons, the security apparatus, and industry. We see John Walsh, whose expertise consists solely of his experience hosting a popular crime-fighting television show, now contributes to security policy through interviews with President Obama in which he makes recommendations about biometric expansion. As the industry continues to search for markets for their products, we see the permeable boundary between media representations and real life as well as the mutually constitutive relationship between marketing and surveillance. These examples also touch on the primary theme of this book: biometric failure. With this new search technology, bodies are imagined as stable entities that can reliably give us definitive proof of identity, creating processes of social stratification in which "material and technological infrastructures divide populations" by gender, race, class, and other axes of identity (Monahan 2010:83). Yet biometric mismatches due to mechanical failures and the technology's inability to work objectively dispute such stability.

In *To Engineer Is Human* the engineering scientist Henry Petroski (1992) explains the role of failure in design. In his examination of a series of technical breakdowns, he repeatedly warns his audience that engineers' failing to think creatively about design errors can lead to devastating accidents. We have an impoverished language for thinking more broadly about such technological failure. We must also think of the intensification of existing inequalities as failures. For example, biometric technologies that rely upon erroneous assumptions about the biological nature of race, gender, and sexuality produce unbiometrifiable bodies, resulting in individuals who are denied their basic human rights to mobility, employment, food, and housing. Although biometric scientists often speak of "false accept" or "false reject" biometric errors, we lack language for thinking about the failures of biometric technologies to contribute to substantive equality.

Marianne Paget (1993), a sociologist of medicine, argues that "clinical medicine is an error-laden activity" and that we lack words to think about medical mistakes in a complex way. Instead language taken from legal and insurance discourses about fault and blame dominates our current ways of speaking about medical error; thus we use the terms *medical malpractice*, *misdiagnosis*, and *negligence*, words that do not get at the affective or emotional nature of medical "mistakes," a word seldom used and considered imprecise for the purpose of insurance and legal claims. Biometrics uses the scientific language of "false acceptance rates" and "failure to enroll rates" to describe biometric failures. These failures are often understood as exceptional, the result of "a few bad apples," rather than as endemic to the science of identification itself. Given the high stakes of biometric identification, our desire for an error-free science of recognition is great. Yet desire to be free from mistakes does not make it so. Paget refers to medical errors as "complex sorrows" to help us think about both their scientific and their emotional implications. Given the devastating consequences of biometric errors for human lives, we need a language that is not restricted to technical terms. Biometric failures as complex sorrows is a beginning.

Examining the multiple failures of biometric technologies, we might ask, "Why do they matter?" Although we are repeatedly assured by industry representatives and government officials that adding biometric technologies to state institutions will replace systemic bias with scientific neutrality, in fact the stakes are very different. In making tremendous profits for security companies, biometrics are part of a larger project that Torin Monahan (2010:37) describes as a realignment of "national security interests with the profit motives of private companies." Examining the scientific assumptions upon which biometric technologies rely as well as the cultural context that drove their development, we see the ways that biometric technologies are deployed in order to deprive people of their most basic human rights. Whether it is by serving up the bodies of poor people of color to a prison industrial complex hungry for increased profits, or whether biometric additions to the border result in some bodies being held static because their fingerprints are "too fine" for the scanner, there are significant consequences to the addition of biometric technologies to state programs. As we saw in the case of those recipients struck from the welfare rolls, deprived of even the most basic means of subsistence following the integration of biometric technologies into social service provision, biometric failures may be a matter of life and death. September 11 created what Paula Treichler (1999:1) calls an "epidemic of signification." The Bush administration quickly moved to try to contain the meaning of September 11 to "an attack on U.S. freedoms," "a terrorist act," and "an act of war." And yet September 11 continually defied such an attempt at containment narratives. From the scientific to the science fictional, the attacks on the Pentagon and Twin Towers have taken on multiple meanings: as a security breach, a result of U.S. foreign policy in the Middle East, an inside job by the Bush administration to give meaning to the Bush presidency, the opportunity to perfect a new security technology, the opportunity to secure American access to Iraqi oil, the opportunity to wage war in Afghanistan, and a war for women's rights. As the tide of explanations threatens to overwhelm, biometric technologies arrive on the scene to contain security breaches, to solve threats with technology, to replace dangerous bodies with zeroes and ones. In flattening three-dimensional bodies into two-dimensional binary code, biometric technologies are part of what Rachel Hall (2009) calls an "aesthetics of transparency," part of the visual culture of post-9/11 security which aims to render complex phenomena transparent, such that there "would no longer be any secrets or interiors, human or geographical, in which our enemies (or the enemy within) might find refuge." One of the central aims of this book has been to investigate what attempts to locate identity in the body tell us about continued contemporary desires for security, for stable narratives, stable borders, and stable bodies that can be made both visible and knowable, whether on the part of state institutions, the security-industrial complex, scientists, or individual travelers who wish to be soothed by security theater. Problematizing the straightforward assertion that biometrics are simply a scientific evolution improving upon older identification technologies, *When Biometrics Fail* instead follows surveillance studies theorists like Torin Monahan in examining security as a social construction. I showed that it is no accident that biometric technologies are a central feature of the post-9/11 security environment. Biometric representations of the body act as a containment narrative, reducing diseased and dying, dangerous and damaged alike to a comforting series of ones and zeroes.

Biometrics are the latest in a long tradition of identification technologies, from anthropometry to the Human Genome Project, that claim to be able to reduce the body to code. All of these earlier technologies paved the way for understanding biometric renderings of the body as bodily truths. Yet, as noted in the epigraph that begins this conclusion, there are attendant risks associated with reducing bodily complexity to binary code. From the material risks of losing welfare benefits due to the unbiometrifiability of the body, to the perils of using biometric technologies to turn prison inmates into numbers to be counted rather than "bodies that matter" (Butler 1993), to their use to mask the state scrutiny of othered bodies at the border, the consequences of biometric failures are significant. As some of the state's most vulnerable bodies are spun into gold through the addition of the biometric scanner, transforming bodies into their component parts usefully eases the passage of marginalized bodies to market.

Biometric failures are numerous. To investigate the relationship between biometric technologies and the cultural context that gave them birth I relied upon the process of what Haraway and Goodeve (2000:115) calls "denaturalization . . . when what is taken for granted can no longer be taken for granted precisely because there is a glitch in the system." Although the most egregious examples of technological breakdown occur reliably where gender, race, class, sexuality, and disability identities are constructed as other, these breakdowns represent only the tip of the iceberg. We have seen a range of biometric failures, from high rates of unbiometrifiability to the misapplication of statistical techniques in the codification of bodily identities and the misunderstanding of cultural trends, even with respect to hairstyle and clothing. The inability of the scanner to identify the long-haired metrosexual man and the woman wearing a tie begs the question "Exactly who can these technologies reliably identify?" Although proponents of biometrics regularly supplied a number of alibis for why the scanner failed to work—whether facial hair, clothing, or disability—it became evident, as noted elsewhere (Treichler 1999), that while cultural theorists generally understand the basics of the scientific objects and processes they study, the same cannot be said for scientists' knowledge of cultural theory. Biometric science consistently fails to examine existing literature on the complexity of bodily identities as well as theory on the interpretation of scientific images. As I noted in chapter 1, biometric scientists seem likely to fail even the most basic question asked on *Project Runway*, as "inappropriate" facial hair and stylish glasses frames are rendered suspect to the scanner.

As biometric industry representatives, computer scientists, and government officials wax poetic about biometric possibilities, we are told that these technologies can do anything: Eliminate crime! End racism! Fix poverty! Recently media hype arose concerning the invention of a new biometric technology that scientists claim is able to read individual thoughts. Asserting that these biometric readers can use the unique electrical impulses of your brainwaves as a biometric identifier, a number of studies were published speculating about new biometric futures in thought policing. In a two-page article in *Wired* describing the possibilities of this new "pass-thought" technology, we are provided with a scant two-sentence postscript briefly reminding us that these technologies may yet be years away and, for the moment, remain in the realm of science fiction (Sandhana 2006). I have attempted to excavate the volumes left unsaid in the postscripts following descriptions of endless biometric possibilities. Although we may consistently hear about the potential of these technologies, I show how limited these technologies remain, through a close analysis of what happens when biometrics are applied to everyday human space. In doing so I found that there is little concern, governmental or otherwise, with adding up the human costs of these technologies or accounting for how biometrics fit into the larger geopolitical picture.

In examining the "contests for meaning" of biometric technologies, to use Haraway's (1991) famous phrase, including those waged among industry providers, government individuals, scientists, and engineers, I also explored whose voices were left out of the debate: prisoners, welfare recipients, immigrants and refugees, and other communities subject to these new identification technologies. Treichler (1999:323) reminds us that "meaning is held in place by social as well as natural reality." As the use of biometric scientists to serve as consultants to the science fiction film *Minority Report* demonstrates, rather than mutually exclusive categories, science, technology, and culture are mutually constitutive. Thus, just as we see a renewed scientific interest in biological theories of race (Hammonds 2006), within biometric science we are witnessing a return to the long discredited fields of anthropometry and physiognomy, serving as a reminder that the pursuit of scientific knowledge is always already bound up with culture.

I also examined the ways that the representation of biometrics allowed for forms of state action that would not have been possible without these new identification technologies. It would have been impossible to install signs at border security locations in a post-9/11 world requiring different security clearance entry points for "Caucasians," "Asian Americans," and "African Americans." It was easily possible, however, to install biometric iris scanners. Using biometric technologies to perform mechanical objectivity, we could encode into the technologies themselves assumptions about the usefulness of racial categories, while obscuring that a form of racial profiling was in fact occurring.

Biometrics became useful as a form of what the security technologist Bruce Schneier (2006) has termed "security theatre," or what the border theorists Peter Andreas and Thomas J. Biersteker (2003) call a "politically successful policy failure." Regardless of how often the technologies broke down, worked differentially depending on race and gender, or cost rather than saved the state money, biometrics were used to suggest that *something* was happening. Thus even when biometric technologies failed, they also succeeded. Whether they were used to assert that the state was getting tough on crime, welfare, or terrorism, biometric technologies were offered up as proof that public-private partnerships were working for the public good, whatever the particular good of the moment might be.

Studying biometric technologies offers a window into state-making in the age of security, including the symbiotic relationship between private enterprise and the state, the increase in information sharing and surveillance, the resurgence of biological racialism, the rise of the prison industrial complex, the criminalization of poverty, and the mutually constitutive relationship between science fiction and real life. Several insights may be drawn from this study of the implementation of biometric identification technologies. Primary among them is that, before formulating further policies regulating biometric technologies, it is imperative that we critically interrogate the assumptions upon which these technologies are based, the limits of any technology to address the larger context of inequality, and the complex relationship of these technologies to their existing cultural context. Moreover the conditions under which they are produced needs to be examined further, as does who gets to participate in conversations about their expansion and development and who is notably absent from these discussions. Our thinking about the implementation of these technologies also needs to move beyond overly simplistic political reactions to the issue of the day, whether it is attempts to get tough on crime, welfare, immigration, or terrorism.

Although 9/11 is the event that drove the biometric industry into stable profitability (Feder 2001), expenditures continue to balloon. Following the biometric industry's consolidations and mergers, global biometric revenues are now forecast to grow from $2.1 billion in 2006 to $5.7 billion in 2010 (International Biometric Group 2006). It is estimated that federal spending on identity programs will reach $2.2 billion by 2012 (Strohm 2008). We need to continue to investigate the investments of billions of dollars on technological solutions to social problems. In an article by Charlotte Bunch (1979) titled "Not by Degrees," she asks feminist theorists to consider how we might change what *is* to what *should be*. Instead of understanding biometrics as a naturalized improvement in state security, and so relying upon new technologies to solve the complex social problems of poverty, homophobia, ableism, sexism, and racism, we must discuss how to build a state concerned with substantive forms of equality. One place to start is with public debate. Giving a radio interview about biometric technologies, I was struck when the journalist asked me, "How come we have not heard more about biometric failures?" Why not indeed.

This book is not arguing that we simply need to refine biometric technologies so they function more efficiently and reliably. Rather we need to understand biometric technologies as a map of the body, one that leaves much out and which fails to represent bodily complexity. We may also understand biometric representations of the body as a map of the contemporary social moment, both producing and reflecting its enduring inequalities, prejudices, and competing values, as well as engraved with continued resistance and hope.

We do not need a perfect technology for representing the body—there is no such technology. Rather we need to think critically about how security imperatives and the development of new technologies are trumping a commitment by the state to address poverty and the perpetuation of intersecting forms of discrimination. In speaking about the consequences of corporeal fetishism, Donna Haraway (1997:160) argues, "We need a critical hermeneutics of genetics as a constitutive part of scientific practice more urgently than we need better map resolution for genetic markers." The same is true of biometric technologies. Technological solutions to social problems have tended to take an approach characterized by the prioritization of security over substantive equality, global lockdown over emancipation, and an uncritical "more is better" approach to new technologies. We need the formulation of technological policy based on principles of inclusiveness and which facilitate substantive claims to equality. Otherwise, as we have seen, in offering to redefine social problems as scientific ones, biometric discourse will simply portray old inequities in new ways.

APPENDIX

1936 Frank Burch, ophthalmologist, proposes the idea of using iris patterns for personal ID (National Center for State Courts, 2002)

1949 "Access control" initially listed in the Federal Standard 1037C, *Glossary of Telecommunication Terms*; Sandia National Laboratories, based in New Mexico, begins developing "science-based technologies that support our national security"

1950s FBI begins efforts to store fingerprint information on punch cards for mechanical searching (P. Jones 2006)

1960s FBI sponsors research at the National Bureau of Standards on automated fingerprint systems (Biometric Consortium 1995)

1963 John Fitzmaurice of Baird-Atomics and Joseph Wegstein and Raymond Moore of the U.S. National Bureau of Standards begin experiments with optical recognition of fingerprint patterns (S. A. Cole 2004)

mid-1960s Identimation develops the hand geometry product Identimat (Guevin 2002)

1967 FBI creates National Crime Information Center (NCIC), a national database of information on wanted people (U.S. Department of Justice 2004)

late 1960s–early 1970s General Electric, McDonnell Douglas, Sperry Rand, and the KMS Technology Center investigate the possibilities of holographic imaging for digital fingerprinting, but holography proves too expensive (S. A. Cole 2004); digital techniques for enhancement of images are developed for NASA and found useful when applied to digital fingerprinting (Cherry and Inwinkelried 2006)

1970s	FBI fingerprint record keeping makes big advances with developments in computer technology and optical scanners (A. Jones 2004); some state and local law enforcement agencies begin installing analog search-and-retrieval systems for fingerprints, using microfilm or videotape, which is more affordable than holography (S. A. Cole 2004)
April 1971	Identimation Corp. patents identification system based on coded fingerprints ("Identimation Corp Patents Identification System Based on Coded Fingerprints" 1971)
1971–75	U.S. government commissions Sandia Labs to compare various biometrics technologies for accuracy (Ruggles 1996)
1972	FBI installs a prototype Automated Fingerprint Identification Systems system, using technology built at Cornell and North American Aviation (this is likely the first use of the acronym AFIS) (S. A. Cole 2004)
1975	EyeDentify is founded by Robert "Buzz" Hill (1999) for development of retinal scanning
1976	Identimation is bought by a new company, Identimat; in the mid-1970s Identimat is the first automatic hand-scanning device commercially available, installed as a time clock system at Shearson Hamill, a Wall Street investment firm (Guevin 2002; Sidlauskas and Tamer 2008)
late 1970s	AFIS marketed to local law enforcement agencies under the brand name Printrak (S. A. Cole 2001); until 1987 hundreds of Identimat devices are installed for access control at government offices and private companies such as Western Electric, U.S. Naval Intelligence, and the Department of Energy (Zhang 2006)
1978	Retinal scanner is patented, followed by a working prototype in 1981 (Hill 1999)
1979	FBI begins testing automatic fingerprint searching using AFIS system; these searches become routine by 1983 (S. A. Cole 2004); Identimat is sold to the Wackenhut Corporation (Sidlauskas and Tamer 2008)
early 1980s	U.S. law enforcement agencies begin to implement AFIS systems for fingerprint storage and analysis; the process still requires time-consuming transfers of paper-based fingerprints to digital information (P. Jones 2006)
1981	The first working prototype of retinal ID is built, leading to EyeDentify's system (Hill 1999)

1982	Printrak system is installed in seven places in the U.S., Canada (RCMP), and Brazil; FBI's fingerprint file has grown to over 75 million cards (on more than 21 million people); approximately 3,000 technicians are employed to search new records against existing files (Charlish 1982)
1983	FBI reports that its entire fingerprint file from 1928 on is now digitized and searchable through AFIS system (S. A. Cole 2001)
1984	Early models of retinal scanners cost up to $60,000; by 1988 EyeDentify's scanner sells for ~$7,000 (M. Browne 1988); EyeDentify's Eyedentification 7.5 is the first retinal scan device made for commercial use (QuestBiometrics 2005); Fingermatrix (1999), in business since 1976, has scanners installed for employee access at U.S. Navy facilities, Chase Manhattan Bank, and First National Bank of Chicago; Fingermatrix has sales of less than $1 million so far, mostly to government and banks; the president of Fingermatrix is "betting on the fingerprint" as the solution to major security breaches to drive the industry forward (E. Johnson 1984)
1985	EyeDentify retinal scanner is marketed as a solution for access control of high-security areas; a single unit retails at $11,000 ("A Company Called EyeDentify" 1985)
March 1986	Industry consultants claim that the biometric security industry will be worth as much as $1 billion a year by 1990 (Reuters 1986)
1986	AFIS is utilized by local law enforcement agencies across the U.S.; there are three principal vendors: Printrak (U.S.), NEC (Japan), and Morpho (France) (S. A. Cole 2001); Australia is the first country to adopt national computerized fingerprint imaging for its law enforcement system (Libov 2001); members of Stellar Systems (makers of Identimat) form Recognition Systems, now the leading hand geometry vendor and a division of Ingersoll-Rand (Guevin 2002)
1987	Identimat ceases production after approximately a decade (Sidlauskas and Tamer 2008); two ophthalmologists, Aran Safir and Leonard Flom, patent the idea of using iris patterns for personal ID (first suggested by Burch in 1936) (National Center for State Courts 2002)
July 1988	Fingermatrix is awarded a $500,000 contract for protecting the U.S. Department of Defense's computer data ("Fingermatrix Awarded PRC Contract for DOD Data Security" 1988)
November 1988	EyeDentify sells retinal scanning system to Chicago Housing Authority; forty-seven units will serve as time clocks for 2,000 Housing Authority employees, the largest such biometric ID system in the U.S. at the time ("Breakthrough for Biometrics" 1988)

1988	Utah State Prison uses an EyeDentify retinal scanner (M. Browne 1988); FBI establishes national DNA Index System, linking city, county, state, and federal databases (Arnold 2006)
late 1980s	Inkless, live-scan fingerprinting becomes available (S. A. Cole 2001)
1989	U.S. Congress provides initial funding to INS to develop automated fingerprint ID system, eventually known as IDENT (U.S. Department of Justice 2004); John Daugman creates algorithms for iris recognition, which he patents in 1994 (now owned by Iridian Technologies and the basis for all current iris recognition systems) (National Center for State Courts 2002); Sandia Labs begins comparative study of several biometrics technologies on volunteer subjects (Holmes, Wright, and Maxwell 1991)
1990	Cook County, Illinois, Sheriff's Department begins using retinal scanning to track prisoners; by 1995 has identified 350,000 criminal defendants through retinal scans (Lichanska n.d.); retinal scanners are used in a variety of locations, including access control at the Pentagon and various intelligence agencies, at the *Arkansas Democrat* newspaper in Little Rock for time clocks, for prisoner identification in Dade County, and at a Florida detention center (Rosen 1990)
1990–91	INS and FBI meet to discuss coordination of fingerprint systems (U.S. Department of Justice 2004)
1991	Digital Biometrics and Printrak announce agreement for development of common interfaces between products ("Digital Biometrics, Printrak International in OEM Purchase Pact" 1991); early work toward realizing automated iris scan recognition is carried out at Los Alamos National Laboratories (Wildes 1997)
December 1991	Los Angeles County Police Department pilots electronic fingerprinting system (Adelson 1992)
1991–93	INS pilots AFIS in San Diego Border Patrol sector (U.S. Department of Justice 2004)
1992	Fingermatrix receives order for two live-scan fingerprinting systems for Nevada from the Western Identification Network, which coordinates AFIS for eleven member states ("Fingermatrix Receives Order" 1992); over the previous year about 100 local police stations installed live-scan fingerprinting machines; the largest are Chicago, with 32, and Philadelphia, with 11 (Adelson 1992); a working group referred to as the Biometric Consortium has been meeting to coordinate biometrics in U.S. government applications; the Consortium is officially chartered in December 1995 (Campbell, Alyea and Dunn 1997)
1993	Thirty-nine states and D.C., along with approximately 350 towns, cities, and counties now have an AFIS system in place (Slade 1993)

April 1994	Los Angeles County becomes the first place in the U.S. where parents applying for welfare for their children have to have biometric finger scans (Adelson 1994)
May 1994	Identix and Digital Biometrics report their first profitable quarters (Adelson 1994)
1994	INS implements IDENT system, independent of FBI's IAFIS system (U.S. Department of Justice 2004); Lancaster County Prison in Pennsylvania is the first prison facility to use iris scanning for prisoner ID (QuestBiometrics 2005)
1995	Identix reports profit for second quarter, record revenue and orders ("Identix Reports Profit" 1995); Texas legislature passes a bill allowing biometric fingerprints to be used in the Texas welfare system (Nabors 2003)
1996	Seven states are in various stages of planning or implementing digital fingerprinting systems for welfare recipients (Newcombe 2001); Identix announces New York Police Department's adoption of fingerprinting technologies ("Morpho to Employ Identix" 1996); Iriscan's first prison system is installed in Lancaster County, Pennsylvania (Coleman 1999)
1997	Industry trade show organizers anticipate that biometrics revenues will more than triple, from $16 million in 1996 to $50 million by 1999, and predict that the number of biometric devices will jump from 8,550 in 1996 to more than 50,000 in 1999 (A. Davis 1997); U.S. Department of Defense takes over and provides the funding for National Biometric Test Center at San Jose State University, established by the Biometric Consortium in 1995 with funding from the Federal Highway Administration (Hanke 1999)
May 1998	U.S. Congress advocates integrating the IDENT and IAFIS systems (U.S. Department of Justice 2004)
1998	Biometrics are linked to the computing industry in new ways, with large players like Microsoft getting involved with standardization efforts (Michael 2003); Identix reports that California is adopting its fingerprinting system for access control in all thirty-three state prisons ("Identix to Supply Networked TouchLock II Systems" 1998)
1999	Seven states (Arizona, California, Connecticut, Illinois, Massachusetts, New Jersey, Texas) are using automated fingerprint ID systems to track welfare recipients; Florida, North Carolina, and Pennsylvania have systems pending; Departments of Motor Vehicles in California, Colorado, Georgia, Hawaii, and Texas require fingerprinting for driver's licenses; INS uses biometrics for border control; Federal Highway Administration is considering implementing fin-

	gerprinting of truck drivers (Hanke 1999); FBI's new computerized crime data system, the NCIC 2000, is implemented (U.S. Department of Justice 2004); biometrics sales have increased from $16 million in 1996 to $60 million in 1999 (Beiser 1999)
2000	National Institute of Justice and Department of Defense begin examining criminal justice applications of biometrics (Miles and Cohn 2006)
May 2001	Fifteen states have introduced antibiometrics legislation (Newcombe 2001)
December 2001	September 11 is identified as driving the biometrics industry into stable profitability (Feder 2001)
2002	Five biometrics companies are undergoing growth and high profits; the companies agree that the attacks of 9/11 hastened the growth of the security market (Chen 2003); Middlesex, Massachusetts, prison facility uses facial recognition technology to track workers; Viisage will create a similar system for Cambridge, Massachusetts's courthouse (Bushnell 2002)
May 2005	Hampshire County, Massachusetts, launches the CHILD Project, which uses biometric technologies to track missing children nationwide (BI2 Technologies 2007)
January 2006	National Institute of Justice announces it has been testing the use of biometrics in prisons by tracking movements of prisoners at the Navy Brig in Charleston, South Carolina (Miles and Cohn 2006)
July 2006	Viisage buys Iridian, and a merger of Viisage and Identix is pending (Parker 2006)
2006	L1 Identity Solutions (2009) is formed from the merger of Viisage and Identix; International Biometric Group (2006) releases an industry report for 2006–10, projecting that global biometrics revenues will grow from $2.1 billion in 2006 to $5.7 billion in 2010, driven by large-scale government and private sector initiatives
2008	An industry official predicts that spending on federal identity programs will grow to $2.2 billion by 2012 (Strohm 2008); FBI website boasts that its IAFIS is "the largest biometric database in the world, containing the fingerprints and corresponding criminal history information for more than 55 million subjects" (Strohm 2008)

NOTES

Notes to Imagining Biometric Security

1. The word *failure* is itself an outgrowth of the nineteenth-century application of the language of business to a much broader arena of social activities, where it was used to refer uniquely to an "entrepreneurial fall from grace" (Sandage 2005:11). In the nineteenth century the word came to include other human endeavors in a linguistic shift that paralleled the application of business terms to everyday life. The expression "I feel like a failure" would have had no meaning in 1800. As the historian Scott Sandage argues, this phrase now "comes so naturally we forget it is a figure of speech: the language of business applied to the soul" (5).
2. Corporeal fetishism is Haraway's extension of Marx's commodity fetishism. Haraway (1997: 135) explains that corporeal fetishism produces "interesting mistakes" by mistaking complex, lively interrelationships for fixed things in a process that helps to smooth the passage of bodies and genes to market. However, she amends Marx to add nonhuman actors to the equation of commodity fetishism (143).
3. In *Dark Angel*, for example, the United States has suffered a massive technological failure referred to only as "the pulse," a giant electrical surge that destroyed the nation's infrastructure. Each episode features the genetically enhanced heroine teaming up with her entourage in order to save a series of imperiled individuals.
4. The ways that the media both informs and is informed by public policy are well studied (Gates 2004; Penley 1997; Treichler 1999). One key example is found in the television show *24*, in which U.S. security agents use torture to extract the "truth" out of terrorists in a timely fashion. Torin Monahan (2010), a surveillance studies theorist, shows that "ticking bomb" scenarios of the type found in *24* were repeatedly referenced in the 2008 presidential debates, despite the fact that their real-world occurrence is extremely rare.
5. After writing about medical error for ten years, Marianne Paget herself was misdiagnosed with back pain rather than smooth muscle cancer. This medical mistake resulted in her early death. Describing her terrible prognosis as a result of medical error, Paget (1993: 9) said, "I have not asked 'Why did this happen to me?' I have

thought rather 'Why not me?' Is there any reason why because of my knowledge I might be excluded from the very thing I have explored so closely? Why not me indeed? Error is fundamental in clinical medicine, and, too, it is endemic. It is not exceptional and uncommon. I, like everyone else, or rather most everyone else, see physicians about my health. Of course, I feel the irony of my position acutely."

6. The term transphobia refers to systemic discrimination directed toward transgender/transsexual (TS/TG) people, including trans men and trans women as well as bois, grrls, and those others who do not feel their gender identity fits comfortably into existing categories.
7. Peter Andreas (2005) uses this term to describe the political utility but material failure of the dramatic security measures used at the U.S.-Mexico border.

Notes to Biometric Failure

1. Verification is sometimes referred to as authentification (P. Reid 2004).
2. CANPASS Air is the forerunner of NEXUS-Air. It is primarily targeted to business travelers; frequent travelers between Canada and the United States who enroll in the program may bypass customs. CANPASS is part of the "Canada Border Services Agency (CBSA) program that facilitates efficient and secure entry into Canada for pre-approved, low-risk air travelers" (www.findbiometrics.com). As with NEXUS, CANPASS travelers must pass a more comprehensive security check before they are allowed to enroll. They then have their irises scanned and are able to go directly to the scanner (located at a number of international Canadian airports) rather than wait in the customs line. This is a program gradually being phased out and replaced with NEXUS. Undoubtedly using biometric technologies in CANPASS has played a role in normalizing the use of biometrics for travelers more broadly.
3. The assertion that biometrics can easily tell whether a body part is alive or dead might also be called into question by the wealth of scientific research on determining the point between the two states, as debated in the *Journal of Death Studies* and the *Journal of Near-Death Studies.*
4. To locate these studies I searched the terms "recognition" and "biometrics" within a number of databases, including *IEEE,* JSTOR, and Ebsco. This allowed me to find scientific articles examining biometric recognition of individuals, of which a subset focused on race or gender. I undertook a detailed analysis of twenty-eight scientific papers written between 1988 and 2008. As the large number of sources located indicates, the computerized verification of identity is a major area of research within the field of biometrics.

Notes to The Beginnings of Biometrics

1. Mike Davis coined the phrase "prison industrial complex" in an article published in 1995 describing the growth in the prison industry in California (cited in Sudbury 2005:xxvii). The prison industrial complex describes the assemblage of corporations, government officials, law enforcement, media, and researchers that profit from incarceration (A. Y. Davis 2003). It includes corporate giants such as Wackenhut, the Correctional Services Corporation, and the Corrections Corporation of

America; law enforcement officials and police officers; professional organizations (including the American Bar Association); and government organizations such as the National Institute of Justice.

2. Following Angela Y. Davis (2003, 2005) and Julia Sudbury (2004), I refuse the criminological language of "inmate" and "corrections." Instead I use the terminology used by the prison abolition movement to refute so-called commonsense notions about crime and criminalization.
3. For a detailed timeline, please see the appendix.
4. I'd like to thank Tara Rodgers for discovering the significance of access control to early biometric development.
5. This phrase was used by Dr. Albert Kligman, a scientific researcher, to describe his feelings upon his first visit to Holmesburg prison: "All I saw before me were acres of skin. It was like a farmer seeing a fertile field for the first time" (cited in Hornblum 1998:37). In his book *Acres of Skin* Allen Hornblum describes the testing of products from cosmetics to chemical warfare agents on prisoners, freed from the "constrictions" of informed consent.
6. An AFIS database is simply a computer system that allows for the storage and retrieval of digital fingerprints; the prints themselves may be saved in any format. AFIS also may refer to hybrid analog-digital systems in which a computerized system is used for the storage and comparison of analog fingerprint records. After the 1980s the term AFIS began to refer to live-scan fingerprinting techniques which were wholly digital—from digitally scanning the fingerprints themselves to the process of storing these digital fingerprints in a database, most often a database that was networked. Thus although the technology itself was not thoroughly digital until the introduction of live scanning in the late 1980s, AFIS systems able to digitally search and store fingerprint records were the first biometric fingerprinting technologies.
7. Early biometric companies made little money. A *New York Times* article in 1984 reported that Fingermatrix had made less than $1 million since its inception in 1976 (Johnson 1984).
8. In 2003 INS was subsumed under the Department of Homeland Security.
9. For example, between 1985 and 1995 the number of incarcerated men doubled. During this same period the number of women, women who are disproportionately black, tripled (Isaac, Lockhart, and Williams 2001; U.S. Department of Justice, Bureau of Justice Statistics 1997).
10. Nor is violence meted out equally to all women; representations of women of color as hypersexualized often result in understandings of their sexual assault as a lesser offense than for white women, while making them more vulnerable to its occurrence (A. Y. Davis 2003).

    During a dramatization of the strip search performed for correctional officers at a prison conference, Angela Davis (2003:81) reports, many of the officers began to cry as they watched the violent invasion of the prisoner's bodily cavities by actors dressed as guards. Without the protection of the uniform and removed from the official, sanctioned space of the prison, the strip search clearly resembled rape.

11. This is an interesting assertion, given that the majority of people who step onto state property (a category that includes national parks, playgrounds, and government buildings in addition to prisons) do not expect to relinquish the majority of their rights.
12. Retinal scanning had almost disappeared after its beginnings in the prison system due to difficulties with its use. However, biometric retinal scanning recently made a comeback as a star player in the war in Iraq. Biometric scans are mandatory for all Iraqis returning to Fallujah from other parts of Iraq (A. Jones 2004).

Notes to Criminalizing Poverty

1. Ultimately the biometric industry's profits from welfare were insufficient. For example, although the industry leader, Identix, turned a profit between 1995 and 1998, those profits were relatively small, ranging from $278,000 to $966,000. The "stocks of biometric identification companies have not soared" as a result of welfare provision. "Industry analysts say that contracts arising from changes in welfare systems are not enough to generate stable profits for Identix Inc., Digital Biometrics Inc., and Fingermatrix Inc" (Feder 2001). It is clear from the reports Identix and Viisage filed with the SEC that welfare remains a small component of their overall profits (http://getfilings.com). Viisage and Identix have since merged to form LI Identity Solutions (see appendix).
2. It is important to note that these distinctions between citizens and noncitizens for benefit support are unusual in an international context. The United States is one of the few industrialized countries to make this distinction with respect to benefit support (Fix and Passel 2002).
3. Identix begins each of its filings to the SEC by asserting, "The costs of fraud are estimated to be in the billions of dollars each year in the United States alone," though they give no reference for this figure (http://getfilings.com). Viisage, which before merging with Identix was another of the biometric industry's major players, additionally gave eliminating fraud as its primary objective.
4. Even though the collection of biometric information may not have any impact on an applicant's immigration status, there is good reason to fear that the information will be shared between government agencies in troubling ways (Howle and Hendrickson 2003).
5. These anxieties were not unfounded given new laws stating that immigrants receiving public benefits were not eligible to apply for a green card (Fix and Passel 2002:9).

Notes to Biometrics at the Border

1. The media in the United States are not alone in their dehistoricized, romantic representations of Canada as a pluralist, more tolerant nation than its neighbor to the south. The Canadian series *Due South*, in which an RCMP officer and a U.S. cop team up to fight crime in Chicago, reveals similar themes. Key to the show is the representation of Constable Benton Fraser as an overly polite, earnest, white Canadian dedicated to solving crime, in contrast to his somewhat crude, rude, and lazy American counterpart. In fact most episodes begin with Fraser attempting to lend

his neighbors a helping hand by solving their minor crimes for free, usually despite the protests of his American comrade.

2. The RCMP has since found itself surrounded by controversy and allegations of corruption (May 2007).
3. There is a relationship between the use of biometric technologies to make both bodies and the border newly visible. One significant commonality between the two is that the process of representing a complex, relational entity (such as a body or a border) as a discrete, reified object facilitates the passage of these newly visualized "things" to market.
4. Schornack's attempts to secure the border recently led to an untimely end of his post. When Mr. and Mrs. Leu, who live right on the U.S. side of the border, built a concrete wall costing $15,000 to keep their American dogs from leaving for British Columbia, Schornack intervened. As the Leus' wall extends three feet into the buffer zone between Canada and the United States, Schornack ordered them to take it down. The Leus refused and sought the aid of a conservative group called the Pacific Legal Foundation, which decries government regulation at the expense of individual property rights. At this point the Bush administration stepped in. Schornack received a call from the special assistant to the president for presidential personnel, who asked Schornack to back down, questioned his patriotism, and asked whether he was a Republican. Schornack refused to back down and was fired the following day. Interestingly Schornack is a Republican who campaigned for Bush twice. In fact his ties to the Republicans are of such long standing that his appointment to the IBC was opposed by a Democratic senator who feared he was too partisan. Schornack contested his dismissal (Schiff 2007). Schornack was replaced by David Bernhardt, formerly a solicitor at the Department of Justice (Tizon 2007).
5. In 2006 Boeing was awarded a contract valued at approximately $2 billion to begin constructing a virtual fence along both the southern and northern U.S. borders as part of yet another emerging border initiative. It is expected that this contract will ultimately yield $10 billion for Boeing.
6. In *How Like a Leaf* (Haraway and Nichols Goodeve 2000) Donna Haraway explains that this title is meant to locate us in the twentieth century using the particular constellation of twentieth-century syntax and figuration (such as the @ sign).
7. The packaging of bodily identities as products that may be sold to the consumer state has contributed significantly to the development of biometrics and other technological means of visualizing bodily identity. That is, the use of biometrics is expanding at a time when the technological imaging of bodies has become an increasingly profitable market. This follows the trend identified by Treichler, Cartwright, and Penley (1998:8) in *The Visible Woman,* in which they argue that women's health concerns are being prioritized just as their health needs are being "efficiently reconstituted as a 'market.'"

Notes to Representing Biometrics

1. Special thanks to Helen Kang for bringing this television program to my attention and for helping me to refine this argument.

2. As Lisa Nakamura (2007:118) notes in her book *Digitizing Race*, Paul Gilroy's "crisis of raciology" dominates *Minority Report*. For example, the film raises the specter of anxieties around racial passing, suggesting that biometric technologies might help to alleviate these anxieties about blurred bodies. After having his eyes replaced, Anderton is identified by a virtual Gap employee as Mr. Yakamoto. At this point Anderton replies with both surprise and disgust. Here the replacement of Anderton's eyes with "Asian eyes" is narratively cast as a joke, meant to provide levity in the middle of an intense chase sequence (Nakamura 2007). And yet the disconnect between Anderton's "Asian eyes" and white body suggests that biometric technologies will be able to definitely locate race in each part of the body. Biometric technologies are deemed central to narratives of anxiety about racial passing.
3. Personal communication with Paula Treichler, June 4, 2008.
4. One famous sequence is of Nazi government representatives lining up their stamps and ink pads, and then violently stamping Jewish identity documents, a representation that highlights the potential for violence inherent in any moment of state identification.
5. This image also serves as a reminder that Lee Harvey Oswald, Timothy McVeigh, and other white, male, ex-military domestic assassins who committed terrorist attacks have been buried in post-9/11 discourse that highlights only people of color as threats.
6. This example usefully exposes the contradiction in mainstream news representations of Muslim American and Muslim Canadian women. The veil is often represented as symbolic of the oppression of Muslim women. However, in keeping a group of young women from competing in a tae kwon do meet, the contest's organizers prevented these women from participating in an activity in which they are asserting their physical power and agency. This example helps to demonstrate the message behind deceptively simple assertions such as "We don't want women to cover because we believe in their freedom and equality."
7. I am roughly paraphrasing bell hooks's (1992) phrase "eating the other."
8. Henry Boitel is the editor of *Biometric Bits*, an internationally recognized summary of developments in identity management.
9. This excerpt is taken from an email exchange between Henry Boitel and John Daugman.

BIBLIOGRAPHY

Abhyankar, A., and S. Schuckers. 2006. "Empirical Mode Decomposition Liveness Check in Fingerprint Time Series Captures." Paper presented at Conference on Computer Vision and Pattern Recognition Workshop, June 17–22, Clarkson University, Potsdam, N.Y.

Abramovitz, Mimi. 2000. *Under Attack, Fighting Back: Women and Welfare in the United States*. New York: Monthly Review Press.

Accurate Biometrics. n.d. Available from http://www.accuratebiometrics.com (accessed March 16, 2009).

ACLU. 2003. "Three Cities Offer Latest Proof That Face Recognition Doesn't Work, ACLU Says." Available from ACLU website (accessed June 2, 2007).

AcSys Biometrics Corp. 2007. "Canadian Company Capabilities: AcSys Biometrics Corp." Available from http://strategis.ic.gc.ca (accessed June 26, 2007).

Adams, James. 1991. *Flying Buttresses, Entropy and O-Rings: The World of an Engineer*. Cambridge: Harvard University Press.

Adelson, A. 1992. "Technology; Faster, More Accurate Fingerprint Matching." *New York Times*, October 11.

———. 1994. "Electronic Fingerprinting Is Growing but Pace Can Be Erratic." *New York Times*, May 31.

Aguilar, Julie. 2010. "More Detainers Placed on Immigrants." *Texas Tribune*, June 21.

"Airport Starts Using Iris Screener." 2005. Available from http://www.vivelecanada.ca (accessed April 27, 2007).

Akkad, Omar El. 2007. "Crown Stays Charges against Suspect in Toronto Raids." *Globe and Mail*, February 23.

"Alanco to Acquire TSI, Inc.; Developer of Proprietary Wireless RF Locating and Tracking Technology Utilized in Security Management and Information Systems." 2001. *Business Wire*, December 13.

Alberts, Sheldon. 2006. *Where, Oh Where Has the Border Gone?* Available from http://forum.blueline.ca (accessed April 27, 2007).

Alberts, Sheldon. 2009. "Renewed Anger as McCain Claims Hijackers Came From Canada." *National Post*, April 24.

Aldridge, Bob. 2008. *America in Peril*. Pasadena: Hope Publishing House. Available from http://www.plrc.org/blog/2008/05/18/data-basing-our-biometrics/ (accessed January 21, 2009).

Alloula, Malek. 1986. *The Colonial Harem*. Minneapolis: University of Minnesota Press.

Althusser, Louis. 1971. *Lenin and Philosophy, and Other Essays*. London: New Left Books.

American Association of Physical Anthropologists. 1996. "AAPA Statement on Biological Aspects of Race." *American Journal of Physical Anthropology* 101:569–70.

Amoore, Louise, and Alexandra Hall. 2009. "Taking People Apart: Digitised Dissection and the Body at the Border." *Environment and Planning D: Society and Space* 27:444–64.

Andreas, Peter. 2005. "The Mexicanization of the U.S.-Canada Border: Asymmetric Interdependence in a Changing Security Context." *International Journal* (Spring): 449–62.

Andreas, Peter, and Thomas J. Biersteker. 2003. *The Rebordering of North America: Integration and Exclusion in a New Security Context*. New York: Routledge.

Ansari, Usamah. 2008. " 'Should I Go and Pull Her *Burqa* Off?' Feminist Compulsions, Insider Consent and a *Return to Kandahar*." *Critical Studies in Media Communication* 25, no. 1:48–67.

Antonelli, A., et al. 2006. "Fake Finger Detection by Skin Distortion Analysis." *IEEE Transactions on Information Forensics and Security* 1, no. 3:360–73.

Anzaldúa, Gloria. 1987. *Borderlands/La frontera: The New Mestiza*. San Francisco: Spinsters/Aunt Lute.

Appleby, Timothy, and Unnati Gandhi. 2006. "The Evening When All Hell Broke Loose." *Globe and Mail*, June 6.

Arena, Kelli, and Carol Cratty. 2008. "FBI Wants Palm Prints, Eye Scans, Tattoo Mapping." *CNN.com*, February 4.

Arnold, Bruce. 2006. "Caslon Analytics: Biometrics." Available from http://www.caslon.com.au (accessed May 20, 2009).

Associated Press. 1995. "Connecticut Lawmakers Approve Strict New Welfare Rules." *New York Times*, June 4.

——. 1996. "State to Make Facial Images of Welfare Recipients." *Patriot Ledger*, April 13.

——. 2006. "Canada-U.S. Border Obscure in Some Places." Ctv.ca. Available from http://www.ctv.ca (accessed September 30, 2006).

Austen, Tim. 2008. "Man Guilty in Terror Plot." *New York Times*, September 25.

Bahdi, Reem. 2003. "No Exit: Racial Profiling and Canada's War against Terrorism." *Osgoode Hall Law Journal* 41, nos. 2–3:293–316.

Bannerji, Himani. 2000. *The Dark Side of the Nation: Essays on Multiculturalism, Nationalism, and Gender*. Toronto: Canadian Scholars' Press.

Barlow, Maude. 2005. *Too Close for Comfort: Canada's Future within Fortress North America*. Toronto: McClelland and Stewart.

Barry, Andrew. 2001. *Political Machines: Governing a Technological Society*. New Brunswick, N.J.: Athlone Press.

Barry, Tom. 2010. "Jailing the American Dream." *Utne Reader,* March–April:58–67.

Beckwith, Barry. 1999. "Illinois Department of Human Services." Available from http://www.dss.state.ct.us (accessed March 19, 2009).

Beiser, Vince. 1999. "Biometrics Breaks into Prisons." Available from Wired.com (accessed November 19, 2007).

Bell, Charles. 2010 [1806]. "Essay on the Anatomy of Expression in Painting." Nabu Press.

Bell, Stewart. 2007. "Homegrown Extremism on Rise: CSIS: Briefing to PM Says Parents May Be to Blame." *National Post (Canada)*, March 20.

Bergstein, Brian. 2003. "Biometric Passport to Be Tested at U.S. Border Checkpoints." *The Age*. Available from http://www.theage.com.au (accessed March 4, 2011).

Berkowitz, Bill. 2001. *Prospecting Among the Poor: Welfare Privatization*. Oakland: Applied Research Center.

Bernstein, Nina. 2000. "Experts Doubt New York Plan to Fingerprint for Medicaid." *New York Times*, August 30.

Biemann, Ursula. 2001. "Writing Desire." *Feminist Media Studies* 1, no. 2:251–58.

Bigo, Didier. 2002. "Security and Immigration: Towards a Critique of the Governmentality of Unease." *Alternatives: Global, Local, Political* 27.

Bioidentification. 2007. "Biometrics: Frequently Asked Questions." Available from http://www.bromba.com (accessed April 27, 2007).

Biometric Consortium. 1995. *About Us*. Available from http://www.biometrics.org (accessed November 22, 2007).

Bivins, Thomas H. 1987. "The Body Politic: The Changing Shape of Uncle Sam." *Journalism Quarterly* 64, no. 1:13–20.

Blatchford, Christie. 2006. "Ignoring the Biggest Elephant in the Room." *Globe and Mail*, June 5.

Boa, Krista. 2006. "Biometric Passports: A Response to the Western Hemisphere Travel Initiative?" Available from http://www.anonequity.org (accessed April 27, 2007).

Bowermaster, David. 2007. "Blaine Couple Fight to Retain Backyard Wall Near Canada Border." *Seattle Times*, April 11.

Bowker, Geoffrey C., and Susan Leigh Star. 1999. *Sorting Things Out: Classification and Its Consequences*. Cambridge: MIT Press.

Brady, Brian. 2006. "Muslims Forced to Lift Veil at Airports." December 24. Available from http://news.scotsman.com (accessed July 20, 2007).

Bratich, Jack Z., Jeremy Packer, and Cameron McCarthy. 2003. *Foucault, Cultural Studies, and Governmentality*. Albany: State University of New York Press.

Braun, David. 2003. "How They Found National Geographic's 'Afghan Girl.'" Available from http://news.nationalgeographic.com (accessed April 17, 2008).
Braun, Hans-Joachim. 1992. "Introduction: Symposium on Failed Innovations." *Social Studies of Science* 22:215.
"Breakthrough for Biometrics." 1988. *American Banker*, November 9.
Brophy-Baermann, Michelle, and Andrew Bloeser. 2008. "Stealthy Wealth: The Untold Story of Welfare Privatization." Available from http://www.allacademic.com/ (accessed March 4, 2011).
Brown, DeNeen L., and Ceci Connolly. 2001. "Suspects Entered Easily from Canada: Authorities Scrutinize Border Posts in Maine." *Washington Post*, September 13.
Browne, Malcolm. 1988. "Technology Rises to the Challenge of Clever Intruders." *New York Times*, May 31.
Browne, Simone. 2004. "Getting Carded: Border Control and the Politics of Canada's Permanent Resident Card." *Conference CTRL: Controlling Bodies*. Montreal.
——. 2009. "Getting Carded: Border Control and the Politics of Canada's Permanent Resident Card." *The New Media of Surveillance*, edited by S. Magnet and K. Gates. London: Routledge.
Brunet-Jailly, Emmanuel. 2004. "Comparing Local Cross-Border Relations under the EU and NAFTA." *Canadian-American Public Policy* 58:1–53.
——. 2006. "Security and Border Security Policies: Perimeter or Smart Border? A Comparison of the European-Union and Canadian-American Border Security Regimes." *Journal of Borderlands Studies* 21, no. 1:3–21.
B12 Technologies. 2007. "Massachusetts Becomes First State to Fully Implement Iris Biometric Technology; B12 Technologies System Continues to Expand Across the Nation." *B12 Technologies Press Release*, June 11. Available from http://www.findbiometrics.com (accessed May 20, 2009).
Buderi, Robert. 2005. "Me, Myself, and Eye." *Technology Review* 64–69.
Bunch, Charlotte. 1979. "Not By Degrees: Feminist Theory and Education." *Quest: A Feminist Quarterly* 5, no. 1.
Bush, George W. 2005. "President Addresses Society of Newspaper Editors Convention." *The White House*. Available from http://georgewbush-whitehouse.archives.gov (accessed March 19, 2011).
Business Editors and Technology Writers. 2001. "Alanco to Acquire TSI, Inc.; Developer of Proprietary Wireless RF Locating and Tracking Technology Utilized in Security Management and Information Systems." *Business Wire*. Scottsdale, Arizona.
Bushnell, Davis. 2002. "Jail Will Use the Latest Facial-Recognition Technology." *Boston Globe*, October 31.
Butler, Judith. 1990. *Gender Trouble: Feminism and the Subversion of Identity*. New York: Routledge.
——. 1993. *Bodies That Matter: On the Discursive Limits of "Sex."* New York: Routledge.
CAEFS. 2004. *CAEFS' Fact Sheets*. Available from http://dawn.thot.net (accessed May 10, 2008).
Campbell, Joseph, Lisa Alyea, and Jeffrey Dunn. 1997. "The Biometric Consortium:

Government Applications and Operations." Available from http://www.biometrics.org (accessed March 15, 2009).
"Canada Charges 17 Terror Suspects." 2006. BBC News, June 4.
"Canada's Edmonton Airport Implements Iris-Recognition Screener at Immigration." 2011. *Edmonton Journal*, January 12. Available from http://www.airportbusiness.com (accessed February 28, 2011).
Canadian Bacon. 1995. Video. LC Purchase Collection, Library of Congress.
"Canadian Biometric ID Documents Public Forum." 2006. Available from http://www.ferenbok.com (accessed January 26, 2009).
Canadian Border Services Agency. 2007. "Travel Documents for Crossing the Border." Available from http://cbsa.gc.ca. (accessed June 15, 2007).
——. 2009. "Nexus AIR." Available from http://www.cbsa-asfc.gc.ca (accessed January 21, 2009).
Canadian Council for Refugees. 2006. *Canadian Council for Refugees E-Chronicle* 1, no. 4. Available from http://ccrweb.ca/ (accessed March 4, 2011).
Canadian Department of Citizenship and Immigration, U.S. Immigration and Naturalization Service, and U.S. Department of State. 2003. *Statement of Mutual Understanding on Information Sharing*. Available from http://www.cic.gc.ca (accessed June 13, 2007).
Canadian HIV/AIDS Legal Network. 2007. "Canada's Immigration Policy as It Affects People Living with HIV/AIDS." Available from http://mqhrg.mcgill.ca (accessed March 4, 2011).
Canadian Senate, Committee on National Security and Defense. 2005. "Borderline Insecure: An Interim Report." June.
Carey, James W. 1989. *Communication as Culture: Essays on Media and Society, Media and Popular Culture*. Boston: Unwin Hyman.
Carney, Scott. 2007. "Thumb-Print Banking Takes India." January 19. Available from Wired.com (accessed July 2, 2007).
Carter, Lee. 2006. "Canada Divided as Losses Mount." *BBC News*, September 6.
Casey, Steve. 1998. *Set Phasers on Stun and Other True Tales of Design, Technology and Human Error*. Santa Barbara, Calif.: Aegean.
CATA Alliance. 2007. "Canadian Biometric Group Forms to Grow Sector, Help with Government Procurement." Available from http://www.cata.ca (accessed June 21, 2007).
Cavoukian, Ann. 1999. *Biometrics and Policing: Comments from a Privacy Perspective*. Edited by Information and Privacy Commissioner of Ontario. Available from http://www.ipc.on.ca (accessed March 20, 2011).
——. 2003. "Statement to the House of Commons Standing Committee on Citizenship and Immigration Regarding Privacy Implications of a National Identity Card and Biometric Technology." Available from http://www.ipc.on.ca (accessed March 20, 2011).
Chan, Wendy, and Kiran Mirchandani. 2002. *Crimes of Colour: Racialization and the Criminal Justice System in Canada*. Peterborough, Ontario: Broadview Press.
Charlish, Geoffrey. 1982. "Technology: Computer Matching for Fingerprints." *Financial Times*, May 26.

Chen, Mandy. 2003. "Biometrics Set To Come Into Its Own." Available from http://support.asmag.com (accessed March 7, 2011).

Cherry, Michael, and Edward Inwinkelried. 2006. "A Cautionary Note About Fingerprint
Analysis and Reliance on Digital Technology." *Juricature* 89, no. 6: 334–8.

Childers, D. G., et al. 1988. "Automatic Recognition of Gender by Voice." *IEEE Transactions on Information Forensics and Security* 1:603–6.

Choudhury, Tanzeem. 2000. "History of Face Recognition." Available from http://vismod.media.mit.edu (accessed December 14, 2007).

Chrétien, Jean, and George W. Bush. 2002. "Joint Statement by President George W. Bush and Prime Minister Jean Chrétien on Implementation of the 'Smart Border' Declaration and Action Plan." *Weekly Compilation of Presidential Documents.* September 9, 1523–24.

Chunn, Dorothy, and Shelley Gavigan. 2006. "From Welfare Fraud to Welfare as Fraud: The Criminalization of Poverty." *Criminalizing Women*, edited by G. Balfour and E. Comack. Black Point, Nova Scotia: Fernwood.

Citizenship and Immigration Canada. 2004. "Safe Third Country Agreement Comes into Force Today." Available from http://www.canada-law.com (accessed March 4, 2011).

Clarke, Roger. 2008. "Introduction to Dataveillance and Information Privacy, and Definitions of Terms." Available from http://www.anu.edu.au (accessed May 9, 2008).

Clarkson, Stephen. 2002. *Uncle Sam and Us: Globalization, Neoconservatism, and the Canadian State.* Toronto: University of Toronto Press and Woodrow Wilson Center Press.

Clear Program Guidelines. 2007. "Clear—Safe, Convenient, Secure." Available from http://www.flyclear.com (accessed July 20, 2007).

Clifford, Stephanie. 2008. "Billboards That Look Back." *New York Times*, May 31.

Cole, C. L. 2000. "One Chromosome Too Many?" *The Olympics at the Millennium: Power, Politics, and the Games*, edited by K. Schaffer and S. Smith. New Brunswick, N.J.: Rutgers University Press.

Cole, C. L., and Sasha Mobley. 2005. "American Steroids: Using Race and Gender." *Journal of Sport and Social Issues* 29, no. 1:3–8.

Cole, Simon A. 2001. *Suspect Identities: A History of Fingerprinting and Criminal Identification.* Cambridge: Harvard University Press.

——. 2004. "History of Fingerprint Pattern Recognition." *Automatic Fingerprint Recognition Systems*, edited by N. Ratha and R. Ruud. New York: Springer.

——. 2007. "How Much Justice Can Technology Afford? The Impact of DNA Technology on Equal Criminal Justice." *Science and Public Policy* 34, no. 2:95–107.

Coleman, Stephen. 1999. "Biometrics in Law Enforcement and Crime Prevention: A Report to the Minnesota Legislature." St. Paul, Minn.: Center for Applied Research and Policy Analysis, Metropolitan State University.

——. 2000. "Biometrics: Solving Cases of Mistaken Identity and More." *FBI Law Enforcement Bulletin* 69, no. 6:9–17.

Collins, Patricia Hill. 1990. *Perspectives on Gender*. Vol. 2, *Black Feminist Thought: Knowledge, Consciousness, and the Politics of Empowerment*. Boston: Unwin Hyman.
"A Company Called EyeDentify of Oregon, USA, Has Developed a Security Device Which Works by Scanning the Pattern of Blood Vessels in the Retina of the Eye, and Which It Claims Is Almost Totally Foolproof." 1985. *Financial Times*, November 1.
"Criminals Cutting Off Fingertips to Hide IDs." *The Boston Channel*. Available from http://www.thebostonchannel.com (accessed January 21, 2009).
Critical Resistance. 2000. "Overview: Critical Resistance to the Prison Industrial Complex." *Social Justice* 3, no. 1.
Currie, Elliott. 1998. *Crime and Punishment in America*. New York: Metropolitan Books.
Cutler, Annie. 2008. "Utah School Implements High-Tech Lunch System." Available from ABC4.com (accessed January 15, 2009).
Czitrom, Daniel J. 1982. *Media and the American Mind: From Morse to McLuhan*. Chapel Hill: University of North Carolina Press.
Das, Ravi. 2007. "Implementing a Multi Modal Security Solution at Your Place of Business." Available from http://ravidas.net (accessed June 26, 2007).
———. 2010. "The Application of Biometric Technologies: 'The Afghan Girl—Sharbat Gula.'" Available from http://www.technologyexecutivesclub.com (accessed July 1, 2010).
Daston, Lorraine, and Peter Galison. 1992. "The Image of Objectivity." *Representations* 40:81–128.
Daugman, John, and Imad Malhas. 2004. "Iris-recognition Border Crossing System in the UAE." *International Airport Review*, no. 2.
Davies, L., J. McMullin, W. Avison, and G. Cassidy. 2001. *Social Policy, Gender Inequality, and Poverty*. Ottawa: Status of Women Canada, February.
Davis, Angela Y. 1981. *Women, Race, and Class*. New York: Random House.
———. 1998. *The Angela Y. Davis Reader*, edited by Joy James. Malden, Mass.: Blackwell.
———. 2003. *Are Prisons Obsolete?* New York: Seven Stories Press.
———. 2005. *Abolition Democracy: Beyond Empire, Prisons, and Torture*. New York: Seven Stories Press.
Davis, Ann. 1997. "The Body as Password." *Wired*, July. Available from http://www.wired.com (accessed May 20, 2009).
Davis, Jim. n.d. "Electronic Fingerprinting for General Assistance Recipients in San Francisco." *Computer Professionals for Social Responsibility*. Available from http://www.gocatgo.com (accessed March 15, 2009).
Dawkins, Richard. 2006. *The Selfish Gene*. 30th edition. Oxford: Oxford University Press.
Day, Stockwell. 2007. "Canada's New Government Invests over $430M for Smart, Secure Borders." Available from http://www.cbsa-asfc.gc.ca (accessed June 14, 2007).
Deleuze, Gilles. 1988. *Foucault*, translated by Sean Hand. Minneapolis: University of Minnesota Press.
"Demolition Man." 2005. Available from http://images.google.ca (accessed June 30, 2010).

Der Derian, James. 2001. *Virtuous War: Mapping the Military-Industrial-Media-Entertainment Network*. Boulder, Colo.: Westview Press.

Derr, M., S. Douglas, and L. Pavetti. 2001. "Providing Mental Health Services to TANF Recipients: Program Design Choices and Implementation Challenges in Four States." Department of Health and Human Services.

de Sola, David. 2006. "Government Investigators Smuggled Radioactive Materials Into U.S." *CNN.com*, March 27.

"Digital Biometrics, Printrak International in OEM Purchase Pact." 1991. *Dow Jones News Service*, December 18.

Digital Descriptor Systems, Inc. 2007. "Compu-Sketch Helps Solve Crime: Retell the Tale." Available from http://www.ddsi-cpc.com (accessed May 9, 2008).

Dobbin, Murray. 2003. *The Myth of the Good Corporate Citizen: Canada and Democracy in the Age of Globalization*. 2nd edition. Toronto: J. Lorimer.

Drache, Daniel. 2004. *Borders Matter: Homeland Security and the Search for North America*. Halifax, Nova Scotia: Fernwood.

Dubow, Saul. 1995. *Scientific Racism in Modern South Africa*. Cambridge: Cambridge University Press.

Duster, Troy. 2005. "Race and Reification in Science." *Science* 207, no. 5712:1050–51.

Dyer, Gwynne. 2006. "Make That Terror with a Small 't'." *Kingston Whig-Standard (Ontario)*, June 5.

Dyer, Richard. 1997. *White*. London: Routledge.

Eisenberg, Eric. 2001. "Building a Mystery: Toward a New Theory of Communication and Identity." *Journal of Communication* 51, no. 3:534–52.

Ekman, Paul, Wallace V. Friesen, and Phoebe Ellsworth. 1972. *Emotion in the Human Face: Guide-lines for Research and an Integration of Findings*. New York: Pergamon Press.

Ericson, Richard Victor, and Kevin D. Haggerty. 1997. *Policing the Risk Society*. Toronto: University of Toronto Press.

Eubanks, Virginia. 2006. "Technologies of Citizenship: Surveillance and Political Learning in the Welfare System." *Surveillance and Security: Technological Politics and Power in Everyday Life*, edited by T. Monahan. New York: Routledge.

Faludi, Susan. 2007. *The Terror Dream: Fear and Fantasy in post-9/11 America*. New York: Metropolitan Books.

Farnell, Brenda, and Laura Graham. 1998. "Discourse-Centered Methods." *Handbook of Methods in Cultural Anthropology*, edited by R. Bernard. Walnut Creek, Calif.: Altamira Press.

Fausto-Sterling, Anne. 2000. *Sexing the Body: Gender Politics and the Construction of Sexuality*. New York: Basic Books.

Feder, Barnaby. 2001. "Technology & Media; A Surge in Demand to Use Biometrics." *New York Times*, December 17.

Feeley, Malcolm, and Jonathan Simon. 1994. "Actuarial Justice: The Emerging New Criminal Law." *The Futures of Criminology*, edited by David Nelken. London: Sage.

Fein, E. B. 1995. "A Test for Welfare Fraud Is Expanded to Families." *New York Times*, November 11.

Fellows, Mary Louise, and Sherene Razack. 1998. "The Race to Innocence: Confronting Hierarchical Relations among Women." *Journal of Race, Gender and Justice* 1, no. 2:335–52.

Fingermatrix. 1999. "Annual Report Pursuant to Section 13 or 15(d) of the Securities Exchange Act of 1934." Available from http://google.brand.edgar-online.com (accessed May 20, 2009).

"Fingermatrix Awarded PRC Contract for DOD Data Security." 1988. *PR Newswire*, July 19.

"Fingermatrix Receives Order for Two Live-Scan Fingerprinting Systems." 1992. *Westchester County Business Journal*, January 6.

Firestone, D. 1995. "100,000 New Yorkers May Be Cut Off Welfare in Crackdown." *New York Times*, August 8.

Fish, Jefferson. 2000. "What Anthropology Can Do for Psychology: Facing Physics Envy, Ethnocentrism, and a Belief in 'Race.'" *American Anthropologist* 102, no. 3:552–63.

Fisk, Robert. 2006. "The Case of the Toronto 17: Has Racism Invaded Canada?" *Counterpunch*, June 12.

Fix, Michael, and Jeffrey Passel. 2002. "The Scope and Impact of Welfare Reform's Immigrant Provisions." Available from http://www.aecf.org (accessed February 15, 2009).

Fix, Michael, and Wendy Zimmermann. 1998. "Declining Immigrant Applications for Medi-Cal and Welfare Benefits in Los Angeles County." Available from http://www.urban.org (accessed March 15, 2009).

——. 1999. *All Under One Roof: Mixed-Status Families in an Era of Reform*. Urban Institute.

Fletcher, Meg. 2006. "Will Pilot Program Fly?" *Business Insurance: Industry Focus*, no. 5:20.

Foreign Affairs and International Trade Canada. 2001. "The Canada-U.S. Smart Border Declaration." Available from http://www.dfait-maeci.gc.ca (accessed April 27, 2007).

——. 2004. "Smart Border Action Plan Status Report." Available from http://geo.international.gc.ca (accessed April 27, 2007).

Foucault, Michel. 1965. *Madness and Civilization: A History of Insanity in the Age of Reason*, translated by R. Howard. New York: Vintage Books.

——. 1972. *Archaeology of Knowledge*. New York: Routledge.

——. 1973. *The Birth of the Clinic: An Archaeology of Medical Perception*. New York: Vintage Books.

——. 1975. *Discipline and Punish: The Birth of the Prison*. New York: Vintage Books.

——. 1978. *The History of Sexuality*. New York: Pantheon Books.

——. 1984. *The Foucault Reader*, edited by Paul Rabinow. New York: Pantheon Books.

Foundation for Information Policy Research. 2008. "FIPR Submission to Home Affairs Committee on ID Cards." Available from http://www.fipr.org (accessed February 17, 2008).

Frank, Thomas. 2009. "Lock 'Em Up: Jailing Kids is a Proud American Tradition." *Wall Street Journal*, April 1.

Freed, Jeremy. 2009. "Making Crime Pay." Available from http://www.pbs.org (accessed March 4, 2011).

Freeman, Alan. 2006. "Alienation at Home, Criticism from Abroad: U.S. Politician Blasts 'South Toronto' as a Hotbed of Islamic Extremism." *Globe and Mail*, June 9.

Freeze, Colin. 2008. "Terrorism Laws Pass Their First Test as Youth Convicted in Homegrown Plot." *Globe and Mail*, September 25.

"Frequently Asked Questions: Ending Finger Imaging." 2003. Available from http://www.cfpa.net/ (accessed March 4, 2011).

Friscolanti, Michael. 2007. "The Four-Million Dollar Rat." *Maclean's*, February 12, 20.

Fry, Earl H., and Jared Bybee. 2002. "NAFTA 2002: A Cost/Benefit Analysis for the United States, Canada and Mexico." *Canadian-American Public Policy* 49:1–33.

Gabriel, Christina, and Laura MacDonald. 2004. "Of Border and Business: Canadian Corporate Proposal for North American 'Deep Integration.'" *Studies in Political Economy* 74 (Autumn): 79–100.

Galloway, Gloria. 2008. "Updated System of Iris Scans, Fingerprints to Biometrically Identify Workers." *Globe and Mail*, May 10.

Garland, David. 2001a. *The Culture of Control: Crime and Social Order in Contemporary Society*. Chicago: University of Chicago Press.

———. 2001b. *Mass Imprisonment: Social Causes and Consequences*. Thousand Oaks, Calif.: Sage.

Gates, Kelly. 2004. "Our Biometric Future: The Social Construction of an Emerging Information Technology." Institute of Communications Research, University of Illinois at Urbana-Champaign.

Geoghegan, Tom. 2005. "The Password for Your Next Phone Is . . ." *BBC News Magazine*, June 9.

Gibbs, Frederic A., and Erna L. Gibbs. 1941. *Atlas of Electroencephalography*. Cambridge, Mass.: John Libbey Eurotext.

Gibson, Diana. 2007. "Commandeering the Continent: Military Integration, Big Oil and the Environment." *Integrate This! Challenging the Security and Prosperity Partnership (SPP) of North America*. Ottawa. Available from http://www.canadians.org (accessed March 20, 2011).

Gilens, Martin. 1999. *Why Americans Hate Welfare: Race, Media, and the Politics of Antipoverty Policy*. Chicago: University of Chicago Press.

Gill, Pat. 1997. "Technostalgia: Making the Future Past Perfect." *Camera Obscura: A Journal of Feminism, Culture and Media Studies* 40:163–79.

Gilliom, John. 2001. *Overseers of the Poor: Surveillance, Resistance, and the Limits of Privacy*. Chicago: University of Chicago Press.

Gilmore, Ruth Wilson. 2007. *Golden Gulag: Prisons, Surplus, Crisis, and Opposition in Globalizing California, American Crossroads*. Berkeley: University of California Press.

Givens, G., et al. 2004. "How Features of the Human Face Affect Recognition: A Statistical Comparison of Three Face Recognition Algorithms." Paper presented at Proceedings of the 2004 IEEE Computer Society Conference on Computer Vision and Pattern Recognition.

Glave, James. 1997. "Prisons Aim to Keep, and Keep Ahead of, Convicts." *Wired*, December 1.

Goldstein, Richard. 2001. "Queer on Death Row." *Village Voice*, March 20.

Golomb, B. A., D. T. Lawrence, and T. J. Sejnowski. 1991. "SexNet: A Neural Network Identifies Sex from Human Faces." *Advances in Neural Processing Systems* 572–77.

Gomm, Karen. 2005. "U.K. Agency: Iris Recognition Needs Work." News.com, October 20.

——. 2007. "Finding Business for Biometrics." Available from http://news.zdnet.com (accessed June 21, 2007).

Gooday, Graeme. 1998. "Rewriting the 'Book of Blots': Critical Reflections on Histories of Technological Failure." *History and Technology* 14, no. 4:265–91.

"Google Refuses to Rule Out Face Recognition Technology Despite Privacy Rows." 2010. *Daily Mail*, May 21.

Gorham, Beth. 2006. "Wilson Not Worried about U.S. Troops Guarding Border." *Globe and Mail*, March 24.

Gould, Stephen Jay. 1996. *The Mismeasure of Man*. Revised and expanded edition. New York: Norton.

Govan, Fiona. 2010a. "'Bin Laden' Is Alive and Well and Living in Spain." *Sydney Morning Herald*, June 24.

——. 2010b. "FBI Admits Spanish Politician Was Model for 'High-tech' Osama bin Laden Photo-fit." *Telegraph*, January 16.

Government of Canada. 2004. "The Canada-U.S. Trade and Investment Partnership." Available from http://geo.international.gc.ca (accessed March 22, 2007).

——. 2007. "Did You Know? Facts about Canada and the U.S." Available from http://www.canadianally.com (accessed April 10, 2007).

Gregory, Peter, and Michael Simon. 2008. *Biometrics for Dummies*. Indianapolis: Wiley.

Guevin, Laura. 2002. "It's All About the Applications." *BiometriTech Newsletter*, July 24.

Gutta, Srinivas, Harry Wechsler, and P. J. Phillips. 1998. "Gender and Ethnic Classification of Face Images." *Third IEEE International Conference on Automatic Face and Gender Recognition*. 194–99.

Gutta, Srinivas, et al. 2000. "Mixture of Experts for Classification of Gender, Ethnic Origin and Pose of Human Faces." *IEEE Transactions on Neural Networks* 11, no. 4.

Hall, Rachel. 2009. "Of Ziploc Bags and Black Holes: The Aesthetics of Transparency in the War on Terror." *The New Media of Surveillance*, edited by Shoshana Magnet and Kelly Gates. New York: Routledge.

Hammonds, Evelynn. 2006. "Straw Men and Their Followers: The Return of Biological Race." Available from http://raceandgenomics.ssrc.org (accessed April 21, 2008).

Hanke, Holly. 1999. "Digital Persona. Metroactive." *Metro: Silicon Valley's Weekly Newspaper*, September 9.

Haraway, Donna Jeanne. 1989. *Primate Visions: Gender, Race, and Nature in the World of Modern Science*. New York: Routledge.

——. 1991. *Simians, Cyborgs, and Women: The Reinvention of Nature.* New York: Routledge.
——. 1997. *Modest_Witness@second_Millennium.FemaleMan_Meets_OncoMouse: Feminism and Technoscience.* New York: Routledge.
——. 2003. *The Companion Species Manifesto: Dogs, People, and Significant Otherness.* Chicago: Prickly Paradigm Press.
Haraway, Donna, and Thyrza Nichols Goodeve. 2000. *How Like a Leaf: An Interview with Thyrza Nichols Goodeve.* New York: Routledge.
Harding, Sandra G. 1993. *The "Racial" Economy of Science: Toward a Democratic Future.* Bloomington: Indiana University Press.
Hartney, Christopher. 2006. "U.S. Rates of Incarceration: A Global Perspective." Available from http://www.nccd-crc.org (accessed February 21, 2008).
Hartung, William, and Jennifer Washburn. 1998. "Lockheed Martin: From Warfare to Welfare." *The Nation*, March 2.
Hauppage, L. 1993. "Suffolk Votes Fingerprinting For Welfare." *New York Times*, September 15.
Hays, Ronald J. 1996. "INS Passenger Accelerated Service System (INSPASS)." Available from http://www.biometrics.org (accessed June 29, 2007).
Hill, Robert. 1999. "Retina Identification." *Biometrics: Personal Information in a Networked Society*, edited by Anil Jain, Ruud Bolle, and Sharath Pankanti. Springer.
Hillmer, Norman. 2004. "United States, Canada's Relations with." *The Oxford Companion to Canadian History*, edited by G. Hallowell. Oxford: Oxford University Press.
"Hi-tech 'Threat' to Private Life." 2007. BBC News, March 26.
Ho, D., and P. Bieniasz. 2008. "HIV-1 at 25." *Cell* 133, no. 4:561–65.
Holmes, James, Larry Wright, and Russell Maxwell. 1991. "A Performance Evaluation of Biometric Identification Devices." Sandia Report. SAND91-0276.
hooks, bell. 1992. *Black Looks: Race and Representation.* Boston: South End Press.
Hornblum, Allen. 1998. *Acres of Skin: Human experiments at Holmesburg Prison: A Story of Abuse and Exploitation in the Name of Medical Science.* Routledge.
Howle, E., and S. M. Hendrickson, eds. 2003. "Statewide Fingerprint Imaging System." *California State Auditor.* Available from http://www.bsa.ca.gov (accessed March 20, 2011).
Howse, Robert. 1996. *Workfare: Theory, Evidence and Policy Design.* Toronto: Centre for the Study of State and Market.
Hruska, Joel. 2007. "FBI Planning World's Largest Biometric Database." *Ars Technica*, December 24.
Hsu, Spencer. 2009. "In 'Virtual Fence,' Continuity With Border Effort by Bush." *Washington Post*, May 9.
Hua-ming, Li, Ming-quan Zhou, and Guo-hua Geng. 2004. "Rapid Pose Estimation of Mongolian Faces Using Projective Geometry." *Proceedings of the 33rd Applied Imagery Pattern Recognition Workshop.*
Hufbauer, Gary Clyde, and Gustavo Vega-Canovas. 2003. "Whither NAFTA: A Common Frontier?" *The Rebordering of North America: Integration and Exclusion in a*

New Security Context, edited by P. Andreas and T. J. Biersteker. New York: Routledge.
IAFIS. 2008. "Integrated Automated Fingerprint Identification System or IAFIS: What is it?" Available from http://www.fbi.gov (accessed May 20, 2009).
"Identimation Corp Patents Identification System Based on Coded Fingerprints." 1971. *New York Times Abstracts,* May 1.
"Identix Reports Profit for Second Quarter, Record Revenue, Record Orders." 1995. *Business Wire.*
"Identix to Supply Networked TouchLock II Systems to California Department of Corrections for Improved Staff Security." 1998. *Business Wire,* July 9.
"Illinois Announces AIM System." 1997. *Biometrics in Human Services Users Group* 1, no. 3.
Immigration Policy Center, American Immigration Council. 2009. *Secure Communities: A Fact Sheet.*
"In Depth: Toronto Bomb Plot." 2006. CBC News Online, August 4.
Innis, Harold Adams. 1950. *Empire and Communications.* Oxford: Clarendon Press.
——. 1951. *The Bias of Communication.* Toronto: University of Toronto Press.
Innis, Harold Adams, and Oliver Baty Cunningham Memorial Publication Fund. 1930. *The Fur Trade in Canada: An Introduction to Canadian Economic History.* New Haven: Yale University Press.
International Biometric Group. n.d. "IBG Reports and Research." Available from http://www.biometricgroup.com (accessed May 20, 2009).
——. 2006. "Biometrics Market and Industry Report Addresses Biometric Investment Landscape Through 2010." Available from http://www.biometricgroup.com (accessed May 20, 2009).
"Iris and Retinal Identification." 2007. Available from http://et.wcu.edu (accessed November 22, 2007).
Isaac, Alicia R., Lettie L. Lockhart, and Larry Williams. 2001. "Violence against African American Women in Prisons and Jails: Who's Minding the Shop?" *Journal of Human Behavior in the Social Environment* 4, nos. 2–3:129–53.
"Jail Biometric Glitch 'Limited.'" 2005. BBC News. Available from the BBC website (accessed February 28, 2008).
Jain, Amit, and Jeffrey Huang. 2004. "Integrating Independent Components and Support Vector Machines for Gender Classification." Paper presented at Proceedings of the 17th International Conference on Pattern Recognition.
Jain, Anil, Sarat Dass, and Karthik Nandakumar. 2004a. "Can Soft Biometric Traits Assist User Recognition?" *Proceedings of* SPIE Defense and Security Symposium, edited by Edward Carapezza. Orlando: International Society for Optical Engineering.
——. 2004b. "Soft Biometric Traits for Personal Recognition Systems." *Proceedings of the International Conference on Biometric Authentication,* edited by David Zhang and Anil Jain. Hong Kong: Springer-Verlag.
James, Joy. 2007. "Warfare in the American Homeland: Policing and Prison in a Penal Democracy." Durham: Duke University Press.

James, Royson. 1999. "City Welfare Fingerprint Plan Flops." *Toronto Star*, May 29.
Jay, Martin. 1993. *Downcast Eyes: The Denigration of Vision in Twentieth-century French Thought*. Berkeley: University of California Press.
Johnson, Elaine. 1984. "Planning to Outfox Office Security? This Handy Device Might Finger You." *Wall Street Journal*, July 2.
Jones, Alex. 2004. "Alex Jones Show: Prison Planet." Available from http://www.prisonplanet.com (accessed May 20, 2009).
Jones, Phillip. 2006. "Using Biometric Technology to Advance Law Enforcement." *Forensic Magazine*, August/September.
Jordan, Lara Jakes. 2005. "Arizona 'Minutemen' Look to U.S.-Canada Border." April 26. Available from http://www.mediavillage.net (accessed June 12, 2007).
Justin, Terry. 1997. "Partnering with Law Enforcement." *Children Today* 24, no. 2:24–25.
Kablenet. 2006. "More Accurate on the Eye." April 5. Available from http://www.theregister.co.uk (accessed July 2, 2007).
Kahn, Alfred J., and Sheila B. Kamerman. 1998. *Big Cities in the Welfare Transition*. New York: Cross-National Studies Research Program, Columbia University School of Social Work.
Kaplan, Caren. 2006. "Precision Targets: GPS and the Militarization of U.S. Consumer Identity." *American Studies Association* 58, no 3: 693–713.
Kassamali, Sumayya, and Usamah Ahmad. 2006. "Wounded Sentiments. Multiculturalism, the "Toronto 17," and the National Imaginary." Available from http://toronto.nooneisillegal.org (accessed May 15, 2007).
Kaufman, Jay. 2008. "The Anatomy of a Medical Myth." Available from http://raceandgenomics.ssrc.org (accessed April 21, 2008).
Keisling, Mara. 2007. "Where People and the Surveillance Society Collide." *Conference in Computers, Freedom, Privacy*. Montreal.
Kember, Sarah. 2003. *Cyberfeminism and Artificial Life*. London; New York: Routledge.
Kent, Jonathan. 2005. "Malaysia Car Thieves Steal Finger." BBC News, March 31.
Kessler, Suzanne. 1990. "The Medical Construction of Gender: Case Management of Intersexed Infants." *Signs: Journal of Women in Culture and Society* 15:3–26.
Ketchum, Alton. 1959. *Uncle Sam: The Man and the Legend*. New York: Hill and Wang.
King County Regional Automated Fingerprint Identification System. 2006. "The Future of AFIS Including AFIS Initiatives 2007–2012." May 15. Available from http://www.metrokc.gov (accessed February 18, 2008).
Kiruba, Murali. 2005. "Biometrics." Available from http://ezinearticles.com (accessed February 18, 2008).
Kohler-Hausmann, Julilly. 2007. " 'The Crime of Survival': Fraud Prosecutions, Community Surveillance and the Original 'Welfare Queen.' " *Journal of Social History* Winter:329–54.
Kravets, David. 2010. "Obama Supports DNA Sampling upon Arrest." *Wired.com*, March 10.

Kruckenberg, Kami. 2003. "Frequently Asked Questions about the New Travel Document Requirements." Available from http://www.lawyerintl.com (accessed March 4, 2011).

Kunzel, Regina. 2008. "Lessons in Being Gay: Queer Encounters in Gay and Lesbian Prison Activism." *Radical History Review* 100:11–37.

Lalvani, Suren. 1996. *Photography, Vision, and the Production of Modern Bodies*. Albany: State University of New York Press.

Latour, Bruno, and Steve Woolgar. 1979. *Laboratory Life: The Social Construction of Scientific Facts*. Beverly Hills: Sage.

Le Bon, Gustave. 1960. *The Crowd: A Study of the Popular-Mind*. New York: Viking Press.

Lee, Erika. 2003. *At America's Gates: Chinese Immigration During the Exclusion Era, 1882–1943*. Chapel Hill: University of North Carolina Press.

Leong, Melissa. 2008. "First Guilty Verdict in Toronto 18 Trials." *National Post*, September 25.

Leong, Melissa, and Darryl Konynenbelt. 2007. "Terror Suspects Complain of Cruel Treatment." *National Post*, May 8.

Lettice, John. 2004. "Marine Corps Deploys Fallujah Biometric ID Scheme: 'Lethal Force' Backs Marketing Campaign." *The Register*, December 9.

Levy, Matthys, and Mario Salvadori. 1994. *Why Buildings Fall Down: How Structures Fail*. New York: W.W. Norton.

Lewine, Edward. 2001. "Face-Scan Systems' Use Debated." *St. Petersburg Times*, December 8.

Lewis, Becky. 2003. "Counting with Fingers." *Corrections Today* 102–3.

Li, Yadong, Ardeshir Goshtasby, and Oscar Garcia. 2000. "Detecting and Tracking Human Faces in Videos." *IEEE* 807–10.

Libov, Yevgeniy. 2001. "Biometrics: Technology That Gives You a Password You Can't Share." Available from http://www.sans.org (accessed May 29, 2009).

Lichanska, Agnieszka. n.d. "Retina and Iris Scans." *Espionage Information: Encyclopedia of Espionage, Intelligence, and Security*. Available from http://www.espionageinfo.com (accessed May 20, 2009).

Liddell, C. B. 2007. "Steve McCurry: Capturing the Face of Asia." Available from http://www.culturekiosque.com (accessed July 21, 2007).

Life + Debt. 2001. Video. Distributed by Mongrel Media, Toronto.

Lipartito, Kenneth. 2003. "Picturephone and the Information Age: The Social Meaning of Failure." *Technology and Culture* 44, no. 1:50–81.

Lipowicz, Alice, and Ben Bain. 2008. "Lockheed Wins FBI Contract." *Washington Technology*, December 12.

Little, Margaret. 2001. "A Litmus Test for Democracy: The Impact of Welfare Changes on Single Mothers." *Studies in Political Economy* 66 (Autumn): 9–36.

Little, Margaret J. H. 1998. *No Car, No Radio, No Liquor Permit: The Moral Regulation of Single Mothers in Ontario, 1920–1997*. New York: Oxford University Press.

Littlefield, Melissa. 2005. "Technologies of Truth: The Embodiment of Deception Detection." Department of English, Pennsylvania State University.

Lombroso, Cesare. 1911 [1861]. *Criminal Man, According to the Classification of Cesare Lombroso, briefly summarized by his daughter Gina Lombroso-Ferrero.* New York: Putnam.

L1 Identity Solutions. 2009. Available from http://ir.l1id.com (accessed May 21, 2009).

Longman, Timothy. 2001. "Identity Cards, Ethnic Self-Perception, and Genocide in Rwanda." *Documenting Individual Identity: The Development of State Practices in the Modern World,* edited by J. Caplan and J. C. Torpey. Princeton: Princeton University Press.

Lugo, Alejandro. 2000. "Theorizing Border Inspections." *Cultural Dynamics* 12, no. 3:353–73.

Luibhéid, Eithne. 2002. *Entry Denied: Controlling Sexuality at the Border.* Minneapolis: University of Minnesota Press.

Lynch, Kevin J., and Frank J. Rodgers. n.d. "Development of Integrated Criminal Justice Expert System Applications." Available from http://ai.arizona.edu (accessed November 11, 2007).

Lyon, David. 2001. *Surveillance Society: Monitoring Everyday Life.* Buckingham, England: Open University Press.

———. 2003. *Surveillance as Social Sorting: Privacy, Risk, and Digital Discrimination.* London: Routledge.

———. 2007. *Surveillance Studies: An Overview.* Cambridge: Polity Press.

Lyons, Michael J., et al. 2000. "Classifying Facial Attributes Using a 2-D Gabor Wavelet Representation and Discriminant Analysis." Paper presented at Proceedings of the Fourth IEEE International Conference on Automatic Face and Gesture Recognition.

MacKinnon, Douglas. 2005. "Oh, No, Canada." *Washington Times,* December 16.

Macklin, Audrey. 2001. "Borderline Security." *The Security of Freedom: Essays on Canada's Anti-Terrorism Bill,* edited by R. J. Daniels, P. Macklem, and K. Roach. Toronto: University of Toronto Press.

"Manley and Ridge Release Progress Report on Implementation of Smart Border Action Plan." 2002. Available from http://w01.international.gc.ca (accessed June 14, 2007).

"Maple Leaf Rage." 2006. *Daily Show,* Comedy Central, June.

Marotte, Bertrand. 2007. "Tae Kwon Do Team Knocked Out for Wearing Hijab." *Globe and Mail,* April 16.

Maryland General Assembly. 1998. Public Assistance—Finger Imaging Identification Pilot Program. H.B. 910. Available from http://mgadls.state.md.us (accessed March 16, 2009).

Matsumoto, T., H. Matsumoto, K. Yamada, and S. Hoshino. 2002. "Impact of Artificial Gummy Fingers on Fingerprint Systems." *Proceedings of SPIE Vol. #4677, Optical Security and Counterfeit Deterrence Techniques IV.*

May, Kathryn. 2007. "Tales of Corruption Rock RCMP." *Ottawa Citizen,* March 29.

McCafferty, Dennis. 1997. "States Use Biometrics to Reduce Welfare Fraud." *Washington Technology,* June 12.

McGowan, Kathleen, and Jarrett Murphy. 1999. "No Job too Big for New Welfare-to-Work Grants." *City Limits*, 201.

McIntire, Michael. 2010. "Ensnared by Error on Growing U.S. Watch List." *New York Times*, April 6.

McLarin, Kimberly J. 1995a. "Inkless Fingerprinting Starts for New York City Welfare." Available from http://query.nytimes.com (accessed April 13, 2007).

——. 1995b. "Welfare Fingerprinting Finds Most People Are Telling Truth." Available from http://query.nytimes.com (accessed April 13, 2007).

McMenamin, Jennifer. 2008. "U.S. Eyeing County Case." *Baltimore Sun*, February 21.

McQuaig, Linda. 1995. *Shooting the Hippo: Death by Deficit and Other Canadian Myths*. Toronto: Viking.

"Mexico Urges Canada to Help Oppose Border Fence." 2006. Available from CTV.ca (accessed March 30, 2007).

Meyers, Deborah Wallace. 2006. "From Horseback to High-Tech: U.S. Border Enforcement." Available from http://www.migrationinformation.org (accessed March 26, 2007).

Michael, Katine. 2003. "The Technological Trajectory of the Automatic Identification Industry." School of Information Technology and Computer Science, University of Wollongong.

Miles, Christopher, and Jeffrey Cohn. 2006. "Tracking Prisoners in Jail With Biometrics: An Experiment in a Navy Brig." *National Institute of Justice: NIJ Journal* 253.

Milmo, Cahal. 2007. "Fury at DNA Pioneer's Theory: Africans Are Less Intelligent Than Westerners." *The Independent*, October 17.

Mirchandani, Kiran, and Wendy Chan. 2005. *The Racialized Impact of Welfare Fraud Control in British Columbia and Ontario*. Toronto: Canadian Race Relations Foundation.

——. 2008. *Criminalizing Race, Criminalizing Poverty: Welfare Fraud Enforcement in Canada*. Fernwood.

Mnookin, Jennifer L. 2001. "Scripting Expertise: The History of Handwriting Identification Evidence and the Judicial Construction of Reliability." *Virginia Law Review* 87, no. 8:1723–845.

Moallem, Minoo. 2005. *Between Warrior Brother and Veiled Sister: Islamic Fundamentalism and the Politics of Patriarchy in Iran*. Berkeley: University of California Press.

Molina Guzmán, Isabel. 2006. "Gendering Latinidad through the Elian News Discourse about Cuban Women." *Latino Studies* 3, no. 2:179–204.

Monahan, Torin. 2010. *Surveillance in the Time of Insecurity*. New Brunswick, N.J.: Rutgers University Press.

Monsebraaten, Laurie. 1996. "Fingerprints for Welfare Will 'Frighten' Mentally Ill." *Toronto Star*, May 9.

Montgomery, Cliff. 2001. "Making Crime Pay." *3am Magazine*. Available from http://www.3ammagazine.com (accessed May 20, 2009).

Monture-Angus, Patricia. 1999. *Journeying Forward: Dreaming First Nations' Independence*. Halifax, Nova Scotia: Fernwood.

Morgan, Daniel, and William Krouse. 2005. "CRS Report for Congress—Biometric Identifiers and Border Security: 9/11 Commission Recommendations and Related Issues." Available from http://fpc.state.gov (accessed May 20, 2009).
"Morpho to Employ Identix TP-600 Live-Scan Exclusively In 73-Precinct ID System For NYPD; Largest Single Urban Law Enforcement System; Contract Value Exceeds $4 Million" 1996. *Business Wire*, April 26.
Mosher, Janet, et al. 2004. "Walking on Eggshells: Abused Women's Experience of Ontario's Welfare System." Available from http://dawn.thot.net (accessed May 29, 2009).
Moss, Frank. 2005. "Proposed Western Hemisphere Passport Rules: Impact on Trade and Tourism." United States Senate Committee on the Judiciary. Available from http://www.epic.org (accessed May 31, 2009).
"Move Over Google Maps, Get Ready for 'Google Body.'" 2007. NBC News. Available from http://www.nbc11.com (accessed March 1, 2008).
Muller, Benjamin. Forthcoming. "Travellers, Borders, Dangers: Locating the Political at the Biometric Border." *Politics at the Airport: Spaces, Mobilities, Controls*, edited by M. Salter. Minneapolis: University of Minnesota Press.
Mullins, Justin. 2007. "Digit-Saving Biometrics." Available from http://www.newscientist.com (accessed July 2, 2007).
Mulvey, Laura. 1978. "Visual Pleasure and Narrative Cinema." *Screen* 16, no. 3: 6–18.
Murray, Heather. 2007. "Monstrous Play in Negative Spaces: Illegible Bodies and the Cultural Construction of Biometric Technology." *The Communication Review* 10:347–365.
——. 2009. "Monstrous Play in Negative Spaces: Illegible Bodies and the Cultural Construction of Biometric Technology." *The New Media of Surveillance*, edited by S. Magnet and K. Gates. London: Routledge.
Myser, Michael. 2007. "The Hard Sell." March 15. Available from http://money.cnn.com. (accessed May 29, 2009).
Nabors, Mary Scott. 2003. "Fingering Opportunities in Biometric Technology." *Austin Business Journal*, October 31.
Nakamura, Lisa. 2002. *Cybertypes: Race, Ethnicity, and Identity on the Internet*. New York: Routledge.
——. 2007. *Digitizing Race: Visual Cultures of the Internet*. Minneapolis: University of Minnesota Press.
——. 2009. "Interfaces of Identity: Oriental Traitors and Telematic Profiling in 24." *Camera Obscura* 24, no. 1: 109–133.
Nanavati, Samir, Michael Thieme, and Raj Nanavati. 2002. *Biometrics: Identity Verification in a Networked World*. New York: John Wiley and Sons.
National Center for State Courts. 2002. "Individual Biometrics: Iris Scan." Available from http://ctl.ncsc.dni.us (accessed November 22, 2007).
National Commission on Terrorist Attacks upon the United States. 2004. *The 9/11 Commission Report: Final Report of the National Commission on Terrorist Attacks upon the United States*. New York: W. W. Norton.
National Science and Technology Council Subcommittee on Biometrics. 2006.

"Biometrics History." Available from http://www.biometrics.gov (accessed November 19, 2007).
Nawrot, Richard. 1997. "New York Update." *Biometrics in Human Services Users Group* 1, no. 3.
Neri, Geoffrey A. 2005. "Of Mongrels and Men: The Shared Ideology of Anti-Miscegenation Law, Chinese Exclusion, and Contemporary American Neo-Nativism." *Bepress Legal Series,* Paper 458.
Newcombe, Tod. 2001. "Biometric Breakdown." *Government Technology: Solutions for State and Local Government in the Information Age,* May 6.
Newman, Cathy. 2002. "A Life Revealed." Available from http://magma.nationalgeographic.com (accessed July 21, 2007).
"Newt Gingrich Sorry for Comments about Canada." 2005. CBC News. April 21. Available from http://www.cbc.ca (accessed April 11, 2007).
"A New U.S. Biometrics Agency Created to Manage DOD-wide Responsibilities." 2010. *Homeland Security Newswire,* March 29.
Ngai, Mae M. 2004. *Impossible Subjects: Illegal Aliens and the Making of Modern America.* Princeton: Princeton University Press.
Nickerson, Colin, and Ellen Barry. 2001. "Attack Aftermath: Looking for Answers to the Hunt." *Boston Globe,* September 13.
Norquay, John. 2007. "Differences in Refugee Protection in the United States and Canada." Ottawa, March 23.
Nuñez-Neto, Blas. 2005. "Border Security: The Role of the U.S. Border Patrol." Available from http://www.fas.org (accessed March 27, 2007).
Oates, John. 2006. "Pre-op Transsexuals Favoured with Twin IDS." Available from http://www.theregister.co.uk (accessed July 19, 2007).
Obomsawin, Alanis. 2006. "Waban-aki: People from Where the Sun Rises." Canada. Available from http://www.onf-nfb.gc.ca.
O'Grady, William. 2007. *Crime in Canadian Context: Debates and Controversies.* Don Mills, Ontario: Oxford University Press.
O'Harrow, Robert. 2005. *No Place to Hide.* New York: Free Press.
Olsen, Stephanie. 2002. "Biometric Devices, Information Systems, and Ethical Issues." *CNet News.com,* January 3.
Onley, Dawn. 2004. "Biometrics on the Front Line." *Government Computer News,* August 16.
Ono, Kent A., and John M. Sloop. 2002. *Shifting Borders: Rhetoric, Immigration, and California's Proposition 187.* Philadelphia: Temple University Press.
Ou, Yongsheng, et al. 2005. "A Real Time Race Classification System." Paper presented at Proceedings of the 2005 IEEE: International Conference on Information Acquisition, June 27–July 3, Hong Kong and Macau, China.
Paget, Marianne. 1993. *The Unity of Mistakes.* Philadelphia: Temple University Press.
Paget, Marianne A., and Marjorie L. DeVault. 1993. *A Complex Sorrow: Reflections on Cancer and an Abbreviated Life.* Philadelphia: Temple University Press.
Parker, Akweli. 2006. " New Jersey Eye-scanning Firm Sold for $35 million." *Philadelphia Inquirer,* July 18.

Parthasaradhi, S. T. V., et al. 2005. "Time-Series Detection of Perspiration as a Liveness Test in Fingerprint Devices." *IEEE Transactions on Systems, Man and Cybernetics, Part C: Applications and Reviews* 35, no. 3:335–43.

"Peddling Welfare-Privatization Boondoggles." 2007. Available from http://info.tpj.org (accessed March 16).

Penley, Constance. 1997. *NASA/TREK: Popular Science and Sex in America*. New York: Verso.

Pepe, Michael. 2000. "Buzz about Biometrics." *CRN*, November 27:81.

Perera, Suvendrini. 2003. "The Impossible Refugee of Western Desire." *Lines* 2, no. 3. Available from http://www.lines-magazine.org (accessed May 25, 2009).

Perkel, Colin. 2008. "Spotlight Shone on Canada's New Anti-terror Law." Available from http://www.prosecutable.com (accessed March 11, 2011).

Peters, John Durham. 1999. *Speaking into the Air: A History of the Idea of Communication*. Chicago: University of Chicago Press.

Petroski, Henry. 1992. *To Engineer Is Human: The Role of Failure in Successful Design*. New York: Vintage Books.

——. 1994. *Design Paradigms: Case Histories of Error and Judgment in Engineering*. Cambridge: Cambridge University Press.

Phillips, Lynn. 1994. "Safety Net Performs Vanishing Act: From Welfare to Un-fare? That's How It Will Go Unless Feminists Rally behind Better Reforms." *On the Issues* 3, no. 4:37–41.

Phillips, P. J., et al. 2007. "FRVT 2006 and ICE 2006 Large-Scale Results." *NISTIR 7408: National Institute of Standards and Technology*. Gaithersburg, Md., March.

Polit, D., A. London, and J. Martinez. 2001. "The Health of Poor Urban Women: Findings From the Project on Devolution and Urban Change." Available from http://www.mdrc.org/ (accessed March 10, 2011).

Porter, Bill. 2007. "Lack of Funding Blurs Border Between US, Canada." *Boston Globe*, April 21.

Posel, Deborah. 2001. "Race as Common Sense: Racial Classification in Twentieth-Century South Africa." *African Studies Review* 44, no. 2:87–113.

Poster, Mark. 2001. *The Information Subject*. Amsterdam: OPA.

——. 2007. "The Secret Self: The Case of Identity Theft." *Cultural Studies* 21, no. 1:118–40.

Powell, Michael. 1941. *49th Parallel*. England.

Precious, Tom. 1992. "Pilot Program Uses Fingerprinting to Detect Welfare Fraud." *Times Union*, December 20.

Precise Biometrics. 2006. "Precise Biometrics Knows Who You Are." Available from http://www.oresundit.com (accessed March 1, 2008).

Preston, Richard Arthur. 1977. *The Defence of the Undefended Border: Planning for War in North America, 1867–1939*. Montreal: McGill-Queen's University Press.

Price, Jenny. 2005. "Biometrics: Tomorrow's Technology Today." Available from http://www.csg.org/ (accessed March 10, 2011).

Pugliese, Joseph. 2003. "The Locus of the Non: The Racial Fault-line of 'Middle-Eastern Appearance.'" *borderlands e-journal* 2, no. 3.

———. 2005. "*In Silico Race* and the Heteronomy of Biometric Proxies: Biometrics in the Context of Civilian Life, Border Security and Counter-Terrorism Laws." *Australian Feminist Law Journal* 23:1–32.

QuestBiometrics. 2005. "Biometric Iris Scanning: Using Eyes to Identify." Available from http://www.questbiometrics.com (accessed May 20, 2009).

Raphael, J. 2000. *Saving Bernice: Battered Women, Welfare, and Poverty*. Boston: Northeastern University Press.

Raphael, J., and S Haennicke. 1999. "Keeping Battered Women Safe through the Welfare-to-Work Journey: How Are We Doing?" *Family Violence Option Report*. Chicago: Center for Impact Research.

Razack, Sherene. 2002. *Race, Space, and the Law: Unmapping a White Settler Society*. Toronto: Between the Lines.

———. 2004. *Dark Threats and White Knights: The Somalia Affair, Peacekeeping, and the New Imperialism*. Toronto: University of Toronto Press.

———. 2008. *Casting Out: The Eviction of Muslims from Western Law and Politics*. Toronto: University of Toronto Press.

Reedman, Clive. n.d. "Biometrics and Law Enforcement." Available from http://www.biometrie-online.net (accessed May 20, 2009).

Reese, Ellen. 2005. *Backlash Against Welfare Mothers: Past and Present*. Berkeley: University of California Press.

Reid, John. 2007. "Question the Home Secretary." Available from http://www.number-10.gov.uk (accessed May 9, 2008).

Reid, Paul. 2004. *Biometrics for Network Security*. Upper Saddle River, N.J.: Prentice Hall.

Reuters. 1986. "Time Clocks That Can't Be Tricked." *San Francisco Chronicle*, March 6.

———. 2006a. "Bush Signs Bill Paying for New U.S. Border Fence." MSNBC, October 4.

———. 2006b. "U.S.: Border Security Contract Goes to Boeing." September 22. Available from http://www.corpwatch.org (accessed March 26, 2007).

Richie, Beth. 2005. "Queering Antiprison Work: African American Lesbians in the Juvenile Justice System." *Global Lockdown: Race, Gender, and the Prison-Industrial Complex*, edited by Julia Sudbury. New York: Routledge. 73–86.

Richtel, Matt. 1999. "Access to Lots of Gadgets Could Be at Your Fingertips." *New York Times*, March 4.

Ritter, Jim. 1995. "Eye Scans Help Sheriff Keep Suspects in Sight." *Chicago Sun*, June 22.

Roach, Kent. 2003. *September 11: Consequences for Canada*. Montreal: McGill-Queen's University Press.

Roberts, Dorothy E. 1997. *Killing the Black Body: Race, Reproduction, and the Meaning of Liberty*. New York: Pantheon Books.

———. 2001. "Criminal Justice and Black Families: The Collateral Damage of Over-enforcement." *UC Davis Law Review* 34, no. 4.

Roberts, John W. 1995. "Yesterday and Tomorrow: Prison Technology in 1900 and 2000." *Corrections Today* 57, no. 4.

Robertson, Colin. 2007. "Home Washington Secretariat John Gormley Interview with Colin Robertson." April 11. Available from http://geo.international.gc.ca (accessed March 25, 2009).

Robertson, Craig Murray. 2004. "Passport Please: The U.S. Passport and the Documentation of Individual Identity 1845–1930." Institute of Communication Research, University of Illinois, Urbana-Champaign.

Rogers-Dillon, Robin. 2004. *The Welfare Experiments: Politics and Policy Evaluation.* Stanford: Stanford Law and Politics.

Rosen, D., et al. 2003. "Psychiatric Disorders and Substance Dependence Among Unmarried Low-income Mothers." *Health and Social Work* 28, no. 2: 157–65.

Rosen, Jerome. 1990. "Biometric Systems Open the Door." *Mechanical Engineering-CIME,* November 1.

Rosenberg, Emily S., Dennis Merrill, Lawrence S. Kaplan, Paul S. Boyer, T. Christopher Jespersen, and Mark T. Gilderhus. 2001. "Foreign Relations." *Oxford Reference Online,* edited by P. S. Boyer. Oxford: Oxford University Press.

Rosenblatt, Dana. 2008. "Behavioral Screening: The Future of Airport Security." *CNN.com,* December 17.

Roundpoint, Russell, and Chief Administrative Officer, Mohawk Council of Akwasasne. 2007. "Where People and the Surveillance Society Collide." *Computers, Freedom, Privacy.* Montreal.

Roy, Saumya. 2002. "Biometrics: Security Boon or Busting Privacy?" *PC World,* January 25.

Ruggles, Thomas. 1996. "Comparison of Biometric Techniques." Available from http://www.bioconsulting.com (accessed May 20, 2009).

Ruiz, Maria Victoria. 2005. "Border Narratives, Latino Health and U.S. Media Representation: A Cultural Analysis." Institute of Communications Research, University of Illinois, Urbana-Champaign.

Saeb, M. 2005. "Black Eye for ID Cards." Available from http://www.blink.org.uk (accessed July 2, 2007).

Sallot, J. 2001. "Canadian Connection Suspected in Hijacking." *Globe and Mail,* September 13.

Salter, Mark. 2007. "Canadian post-9/11 Border Policy and Spillover Securitization: Smart, Safe, Sovereign?" *Critical Policy Studies,* edited by M. Orsini and M. Smith. Vancouver: UBC Press.

Salutin, Rick. 2006. "Diverse Till Proven Monolithic." *Globe and Mail,* June 23.

Sandage, Scott A. 2005. *Born Losers: A History of Failure in America.* Cambridge: Harvard University Press.

Sandhana, Lakshmi. 2006. "Your Thoughts Are Your Password." *Wired.com,* April 27. Available from http://www.wired.com (March 11, 2011).

Sandström, Marie. 2004. "Liveness Detection in Fingerprint Recognition Systems." Electrical Engineering, Linköping University.

Schiff, Stacy. 2007. "Politics Starts at the Border." *New York Times,* June 22.

Schiller, Dan. 1999. *Digital Capitalism: Networking the Global Market System.* Cambridge: MIT Press.

——. 2007. *How to Think about Information*. Urbana: University of Illinois Press.

Schlage. 2006. "HandKey Saves Prison Service Money." May 20. Available from http://recognitionsystems.schlage.com (accessed February 21, 2008).

Schneier, Bruce. 2001. "Biometrics in Airports." Available from http://schneier.com (accessed July 2, 2007).

——. 2006. "Last Week's Terrorism Arrests." Available from http://www.schneier .com (accessed March 11, 2008).

Scott, Stefanie. 1996. "Area's Mix is Cited in Welfare Test - Local Urban, Rural Environment Key in Fingerprint Demonstration." *San Antonio Express News*, May 23.

Scott, James C. 1998. *Seeing Like a State: How Certain Schemes to Improve the Human Condition Have Failed*. New Haven: Yale University Press.

"Securing an Open Society: Canada's National Security Policy." 2004. Available from http://www.publicsafety.gc.ca (accessed June 14, 2007).

Shaikh, Siraj, and C.K. Dimitriadis. 2008. "My Fingers Are All Mine: Five Reasons Why Using Biometrics May Not Be a Good Idea." *IEEE Xplore*. Available from ieeexplore.ieee.org (accessed March 10, 2011).

Sharp, David. 2006. "U.S.-Canada Border Security Choked with Weeds." *Desert News*, October 1.

Sheahan, Matthew. 2004. "A4Vision See Biometrics Growing, Raises Nearly $13M." *Thomson Media Inc.*, September 20.

Shephard, Michelle. 2007. "Hearing Looks at Role of Teens in Terror Case." *Toronto Star*, January 15.

Shohat, Ella. 2006. *Taboo Memories, Diasporic Voices*. Durham: Duke University Press.

Shohat, Ella, and Randy Martin. 2002. "Introduction: 911—A Public Emergency?" *Social Text* 20, no. 3:1–8.

Shrybman, Stephen. 2007. "Commandeering the Continent: Military Integration, Big Oil and the Environment." *Integrate This! Challenging the Security and Prosperity Partnership (SPP) of North America*. Ottawa.

Sidlauskas, David, and Samir Tamer. 2008. "Hand Geometry Recognition." *Handbook of Biometrics*, edited by Anil Jain, Patrick Flynn, and Arun Ross. Springer.

Sighele, Scipio. 1901. *La foule criminelle*. Paris: F. Alcan.

Silver, J. 1995. "Fingerprint Identification System Dusts Benefits Fraud." *Government Computer News* 14, no. 22:12.

Slade, Margot. 1993. "Fingerprint System Extends Arm of the Law." *New York Times*, November 12.

Smith, Elliot. 2004. "Money Fled to Security Sector After 9/11; Anyone Feel Safe?" *USA Today*, July 5.

Smith, Anna Marie. 2007. *Welfare Reform and Sexual Regulation*. New York: Cambridge University Press.

Somerville, Margaret A., and Sarah Wilson. 1998. "Crossing Boundaries: Travel, Immigration Human Rights and AIDS." *McGill Law Journal* 43: 781–834.

Specter, Arlen. 2005. "Securing Electronic Personal Data." April 13.

Spivak, Gayatri. 1988. "Can the Subaltern Speak?" *Marxism and the Interpretation of Culture*, edited by C. Nelson and L. Grossberg. Urbana: University of Illinois Press.

Starr, Paul. 1982. *The Social Transformation of American Medicine.* New York: Basic Books.

"State Lines: Digital Prison Stripes." 2004. *Government Computer News*, January 22.

Steinberg, J. 1993. "Coming Soon: Fingerprints at Many Fingertips." *New York Times*, January 10.

Sternberg, Robert. 2007. "Race and Intelligence: Not a Case of Black and White."

Sterne, Jonathan. 2003. *The Audible Past: Cultural Origins of Sound Reproduction.* Durham: Duke University Press.

Stillar, Glenn F. 1998. *Analyzing Everyday Texts: Discourse, Rhetoric, and Social Perspectives.* Thousand Oaks, Calif.: Sage.

Stockstill, Mason. 2005. "A Much Different Picture at Northern Border: Canada." Available from http://lang.dailybulletin.com (accessed March 26, 2007).

Strohm, Chris. 2008. "Federal Identity Programs Boost Biometrics Market." Available from governmentexecutive.com (accessed May 20, 2009).

Stroz Friedberg. 2010. "Source Code Analysis of gstumbler." Available from http://static.googleusercontent.com (accessed June 22, 2010).

Sturgeon, W. 2004. "Law and Policy Cheat Sheet: Biometrics." Available from http://management.silicon.com (accessed June 28, 2007).

——. 2005. "Law & Policy Cheat Sheet: Biometrics." Available from http://management.silicon.com/ (accessed July 28, 2007).

Sturken, Marita, and Lisa Cartwright. 2001. *Practices of Looking: An Introduction to Visual Culture.* Oxford: Oxford University Press.

Sudbury, Julia. 2004. "From the Point of No Return to the Women's Prison: Writing Contemporary Spaces of Confinement into Diaspora Studies." *Canadian Woman Studies/Les cahiers de la femme* 23, no. 2:154–63.

——. 2005. *Global Lockdown: Race, Gender, and the Prison-Industrial Complex.* New York: Routledge.

Tagg, John. 1988. *The Burden of Representation: Essays on Photographies and Histories.* Amherst: University of Massachusetts Press.

Tanaka, Kenichi, et al. 2004. "Comparison of Racial Effect in Face Identification Systems Based on Eigenface and Gaborjet." Paper presented at SICE Annual Conference, Hokkaido Institute of Technology.

Terry, Jennifer. 1990. "Lesbians under the Medical Gaze: Scientists Search for Remarkable Differences." *Journal of Sex Research* 27, no. 3:317–39.

——. 1999. *An American Obsession: Science, Medicine, and Homosexuality in Modern Society.* Chicago: University of Chicago Press.

——. 2008. "Killer Entertainments." Available from http://www.vectorsjournal.org (accessed March 29, 2008).

Thalheim, Lisa, Jan Krissler, and Peter-Michael Ziegler. 2002. "Body Check: Biometric Access Protection Devices and Their Programs Put to the Test." Available from http://www.heise.de (accessed March 20, 2008).

Thobani, Sunera. 2000. "Nationalizing Canadians: Bordering Immigrant Women in the Late Twentieth Century." *Canadian Journal of Women and the Law* 12:279–312.

——. 2007. *Exalted Subjects: Studies in the Making of Race and Nation in Canada.* Toronto: University of Toronto Press.

Thompson, Elizabeth. 2007. "Federal Court Strikes Down Third-country Refugee Agreement." *National Post,* November 29.

Thompson, John Herd, and Stephen J. Randall. 1994. *Canada and the United States: Ambivalent Allies.* Athens: University of Georgia Press.

Tice, Karen. 1998. *Tales of Wayward Girls and Immoral Women: Case Records and the Professionalization of Social Work.* Chicago: University of Illinois Press.

Tilton, Catherine. 2004. "From Hollywood to the Real World: How Biometrics Will Intersect with Ordinary People's Lives." Available from http://www.biometrics.org (accessed March 20, 2009).

Tishkoff, Sarah, and Kenneth Kidd. 2004. "Implications of Biogeography of Human Populations for 'Race' and Medicine." *Nature Genetics* 36, no. 11: S21–S27.

Tizon, Tomas. 2007. "Northwest Boundary Dispute Sparks Firing—An Official Battling a Couple's 4-foot Wall Is Now Fighting for His Job." *The Nation,* July 15.

Tomes, Nancy. 1998. *The Gospel of Germs: Men, Women, and the Microbe in American Life.* Cambridge: Harvard University Press.

"Tommy Thompson on Welfare and Poverty." 2008. *On the Issues.* Available from http://www.ontheissues.org (accessed February 28, 2010).

Trauring, Mitchell. 1963. "Automatic Comparison of Finger-Ridge Patterns." *Nature* 197:938–40.

Treichler, Paula. 1990. "Feminism, Medicine, and the Meaning of Childbirth." In *Body/Politics: Women and the Discourses of Science,* edited by M. Jacobus, E. F. Keller, and S. Shuttleworth. New York: Routledge.

——. 1999. *How to Have Theory in an Epidemic: Cultural Chronicles of AIDS.* Durham: Duke University Press.

Treichler, Paula A., Lisa Cartwright, and Constance Penley. 1998. *The Visible Woman: Imaging Technologies, Gender, and Science.* New York: New York University Press.

Tu, Thanh Ha, Ingrid Peritz, and Bertrand Marotte. 2007. "Lift Face Veils or Don't Vote, Quebec Tells Muslims." *Globe and Mail,* March 24.

Turner, Alan. 2003. "Biometrics in Corrections: Current and Future Deployment." *Corrections Today,* July.

"$24B Spent on Security in Canada Since 9/11." 2008. CBC News. Available from http://www.cbc.ca (accessed March 24, 2008).

Ueki, Kazuya, et al. 2004. "A Method of Gender Classification by Integrating Facial, Hairstyle, and Clothing Images." Paper presented at Proceedings of the Seventeenth International Conference on Pattern Recognition.

U.K. Biometrics Working Group. 2002. "Biometrics for Identification and Authentication—Advice on Product Selection." Available from http://www.idsysgroup.com (accessed March 21, 2009).

Urbina, Ian. 2009. "Despite Red Flags About Judges, a Kickback Scheme Flourished." *New York Times,* March 27.

U.S. Consulate General in Ciudad Juarez. 2007. "American Citizen Services, General

Information." Available from http://ciudadjuarez.usconsulate.gov (accessed March 23, 2007).
U.S. Customs and Border Protection. 2003a. "Agents Added to U.S.-Canada Border to Enhance Homeland Security." Available from http://www.iwar.org.uk (accessed April 27, 2007).
——. 2003b. "CBP Assigns Additional Border Patrol Agents to Increase Northern Border Security." Available from http://www.cbp.gov (accessed March 26, 2007).
U.S. Department of Homeland Security. 2007. "U.S., Canada Expand Frequent Traveler Program to Montreal's Trudeau International Airport." Available from http://canada.usembassy.gov (accessed June 12, 2007).
U.S. Department of Homeland Security, Bureau of Customs and Border Inspection. 2006. Documents Required for Travelers Departing from or Arriving in the United States at Air Ports-of-Entry from within the Western Hemisphere: Final Rule. 71 226. November 24. Available from http://www.dhs.gov (accessed March 4, 2011).
U.S. Department of Justice. 2000. "Implementation Plan for Integrating INS' IDENT and FBI's IAFIS Fingerprint Data." Available from http://www.justice.gov (accessed March 4, 2011).
——. 2001. "Status of IDENT/IAFIS Integration." December 7. Available from http://www.usdoj.gov (accessed May 29, 2009).
——. 2004. "IDENT/IAFIS: The Batres Case and the Status of the Integration Project." Available from http://www.globalsecurity.org (accessed March 4, 2011).
U.S. Department of Justice, Bureau of Justice Statistics. 1997. *Sourcebook of Criminal Justice Statistics.*
U.S. Department of State. 2004. "Revision of NAFTA Professional Procedures for Mexicans." Available from http://travel.state.gov (accessed March 30, 2007).
——. 2005–2006. "Foreign Entry Requirements 2007." Available from http://travel.state.gov (accessed March 24, 2007).
——. 2007. "Canada: Security Assistance." Available from http://www.state.gov (accessed March 22, 2007).
U.S. Embassy. 2007. "U.S.-Canada Relations: General Information." Available from http://canada.usembassy.gov (accessed March 22, 2007).
U.S. Government Accountability Office. 2002a. "Border Security: Implications of Eliminating the Visa Waiver Program." November.
——. 2002b. "Technology Assessment: Using Biometrics for Border Security. November.
U.S. House of Representatives. 2006. *The Need to Implement WHTI to Protect U.S. Homeland Security. Hearing of the Immigration, Border Security, and Claims Subcommittee of the House Judiciary Committee.* Available from Federal News Service.
U.S. House of Representatives, Committee on International Relations, Subcommittee on Western Hemisphere. 2006. *U.S.–Canada Relations: Statement of David M. Spooner, Assistant Secretary for Import Administration, International Trade Administration, U.S. Department of Commerce.* May 25.
U.S. House of Representatives, Minority Staff of the Committee on Homeland

Security. 2005. *The U.S. Border Patrol: Failure of the Administration to Deliver a Comprehensive Land Border Strategy Leaves Our Nation's Borders Vulnerable.* Available from http://homeland.house.gov (accessed March 17, 2008).

U.S. Legislature of the State of Texas. 1995. An Act relating to the provision of services and other assistance to needy people, including health and human services and assistance in becoming self dependent. H. B. No. 1863. Available from http://www.capitol.state.tx.us (accessed March 17, 2008).

U.S. Senate, Committee on the Judiciary. 1996. Hearings on Criminal Identification Systems.

——. 1997. Hearings on FBI Operations. June 4.

U.S. Senate, Committee on the Judiciary, Subcommittee on Immigration. 1997a. Testimony of William H. Fike, Executive Vice President/Vice Chairman Magna International, before the Senate Judiciary Committee, Hearings on Immigration Exit-Entry Tracking on the U.S.-Canadian Border. Detroit, Michigan. October 14. Available from FDCHeMedia, Inc.

——. 1997b. Testimony of Steven G. Miller, Mayor of the City of Port Huron, before the Senate Judiciary Committee, Hearings on Immigration Exit-Entry Tracking on the U.S.-Canadian Border. Detroit, Michigan. October 14. Available from FDCHeMedia, Inc.

U.S. Social Security Administration, Office of the Inspector General. 2000. "The Social Security Administration Is Pursuing Matching Agreements with New York and Other States Using Biometric Technologies." Available from http://www.ssa.gov (March 4, 2011).

U.S. State of Connecticut Department of Social Services. 1999. Available from http://www.ct.gov/dss/site/default.asp (March 19, 2009).

"UT Austin Researchers Determine Lone Star Image System Project Did Not Prevent Welfare Fraud." 1997. Available from http://www.utexas.edu (accessed March 4, 2011).

van der Ploeg, Irma. 2005. *The Machine-Readable Body: Essays on Biometrics and the Informatization of the Body*. St. Maartenslaan, Netherlands: Shaker.

Wacquant, Loïc J. D. 2009. *Punishing the Poor: The Neoliberal Government of Social Insecurity.* Durham: Duke University Press.

Waldby, Cathy. 2000. *The Visible Human Project: Informatic Bodies and Posthuman Medicine*. London: Routledge.

Walkom, Thomas. 2006. "If These Are Terrorists, They Are Second-Rate." *Toronto Star*, June 7.

Wayman, James. 2004. "Biometrics—Now and Then: The Development of Biometrics over the Last 40 Years." Paper presented at Second International BSI Symposium on Biometrics.

Webb, Jim. 2009. "Why We Must Fix Our Prisons." *Parade Magazine*, March 29.

Webb, Maureen. 2007. *Illusions of Security: Global Surveillance and Democracy in the post-9/11 World.* San Francisco: City Lights Books.

Wegstein, J. 1970. "Automated Fingerprint Identification." *NBS Tech*, August: note 538.

Welch, Michael, and Fatiniyah Turner. 2007. "Private Corrections, Financial Infrastructure, and Transportation: The New Geo-economy of Shipping Prisoners." *Social Justice* 34, no. 3–4: 56–78.

Wells, Spencer. 2002. *The Journey of Man: A Genetic Odyssey*. New York: Random House.

Wente, Margaret. 2006. "Generation Jihad: Angry, Young, Born-Again Believers." *Globe and Mail*, June 6.

Wildes, Richard. 1997. "Iris Recognition: An Emerging Biometric Technology." *Proceedings of the IEEE* 85, no. 9:1348–63.

Wilkins, David H. 2007. "Ambassador Wilkins Discusses U.S.-Canadian Relations." USINFO webchat transcript. March 22. Available from http://canada.usembassy.gov (accessed March 4, 2011).

"Will Profiling Make a Difference?" 2010. Roundtable discussion. *New York Times*, July 3.

Williams, Chris. 2008. "Biometrics Exhibit Blushes Over Email Snafu." *The Register*, August 29.

Wong, D. S., and F. Phillips. 1995. "Weld Wants Fingerprints in Aid Cases." *Boston Globe*, October 18.

Woodward, John D., Nicholas M. Orlans, and Peter T. Higgins. 2003. *Biometrics*. New York: McGraw-Hill/Osborne.

Yang, Zhiguang, Ming Li, and Ai Haizhou. 2006. "An Experimental Study on Automatic Face Gender Classification." *18th International Conference on Pattern Recognition*.

Yin, Lijun, Jingrong Jia, and Joseph Morrissey. 2004. "Towards Race-Related Face Identification: Research on Skin Color Transfer." Paper presented at Proceedings of the Sixth IEEE International Conference on Automatic Face and Gesture Recognition. Available from http://ieeexplore.ieee.org (accessed March 4, 2011).

Zabkiewicz, Denise, and Laura Schmidt. 2007. "Behavioral Health Problems as Barriers to Work: Results from a 6-year Panel Study of Welfare Recipients." *Journal of Behavioral Health Services & Research* 34, no. 2:168–85.

Zaslow, M.J., et al. 2006. "Maternal Depressive Symptoms and Low Literacy as Potential Barriers to Employment in a Sample of Families Receiving Welfare: Are There Two-generational implications?" *Welfare, Work, and Well-Being*, edited by M.C. Lennon. New York: Haworth Medical Press.

Zedlewski, S.R., and D.W. Alderson. 2001. "Before and After Reform: How Have Families on Welfare Changed?" Washington, D.C.: Urban Institute.

Zewail, R., A. Elsafi, M. Saeb, and N. Hamdy. 2004. "Soft and Hard Biometrics Fusion for Improved Identity Verification." *MWSCAS: Midwest Symposium on Circuits and Systems*. Hiroshima: IEEE.

Zhang, David. 2006. "Palmprint Authentication." *International Series on Biometrics*. Springer.

Zorzetto, Alicia. 2006. "Canadian Sovereignty at the Northwest Passage." *The Inventory of Conflict and Environment (ICE): Case Studies*. Available from http://www.american.edu (accessed March 4, 2011).

INDEX

Shoshana Amielle Magnet is an assistant professor in the Institute of Women's Studies / Institut d'études des femmes and the Department of Criminology / Département de criminologie at the University of Ottawa and a co-editor, with Kelly Gates, of *The New Media of Surveillance* (2009).

Library of Congress Cataloging-in-Publication Data
Magnet, Shoshana.
When biometrics fail : gender, race, and the technology of identity / Shoshana Amielle Magnet.
p. cm.
Includes bibliographical references and index.
ISBN 978-0-8223-5123-8 (cloth : alk. paper)
ISBN 978-0-8223-5135-1 (pbk. : alk. paper)
1. Biometric identification. 2. Biometric identification—Political aspects.
3. Biometric identification—Social aspects. I. Title.
TK7882.B56M34 2012
570.1'5195—dc23 2011021954